D1380587

JOHN HULL GRUNDY'S

ARTHROPODS OF MEDICAL IMPORTANCE

edited by

Nicholas R. H. Burgess

Arthropods of Medical Importance

ISBN 0 902068 11 3

Published by
NOBLE BOOKS LIMITED
Watch Cottage, Chilbolton, Hampshire

Printed at the Curwen Press, London

X0674

CONTENTS

FOREWORD

John Hull Grundy studied at King's College, London, and at the Chelsea School of Art, before joining the staff of the Royal College of Art. There his curriculum included colour design, etching and engraving, fine-line block production and printing. A number of anatomical studies were made at the Royal College of Surgeons and at the Orpington War Hospital.

In 1942 he took up a wartime post at the Royal Army Medical College, Millbank, London, where he remained as Lecturer in Entomology when the Second World War ended, until his retirement in 1967. A gifted teacher and writer, he established for himself a considerable reputation in his subject and his entomological drawings are widely known. However, the broader scope of his artistic ability has not perhaps been so fully appreciated, and an exhibition in 1976–77 at the Royal Army Medical College consisted of about 150 diverse examples of his work including anatomical studies, scenes from hospital life, animal drawings and publicity material, as well as many entomological drawings.

Although John Grundy has written extensively, most of his work was intended for distribution to his students and very little had in fact been published. A few years ago I was given a large collection of his manuscripts and drawings, and asked by John and his wife Anne to arrange them in a form suitable for publication. In 1979 a collection of essays and illustrations was published entitled 'Medical Zoology for Travellers', and this present work is a further collection of writings consisting of lecture notes, essays and 'conversations' (questions and answers), most of which were used in the teaching of tropical entomology to Army doctors in the period 1950 to 1967. I have taken the liberty of editing these writings to some extent, both in an effort to update the information and scientific names, and to shape them into a comprehensive

7

reference book. I have also added a short chapter on cockroaches which, until recently, have not been accepted as relevant in the epidemiology of human disease.

This book gives ample evidence of the qualities which proved so valuable to John Grundy's students, myself amongst them, over the years. These include his ability to convey the essential facts of the subject in an interesting and entertaining manner and the enthusiasm with which he did so, his knowledge of the subject, and perhaps above all, his particular and unusual skill as an artist. He is a perfectionist in all aspects of his work, and I have known him tear up an apparently perfect and almost completed drawing which was perhaps inaccurate in some small way, rather than present an imperfect result. It almost broke the heart of many a student to see him erase from the blackboard his most exact and finely executed teaching drawings as if they were insignificant doodles!

The individual quality of John Grundy's illustrations lies in his careful study of the subject matter and an appreciation of its potential movement. Many of the drawings in this book are line-drawings, but there are also examples of his particular technique based on the principle of the half-tone screen, where the darker areas are obtained by dots or lines touching and running together, and the lighter parts by smaller and more isolated dots.

It has been a pleasure to collaborate with John and Anne Grundy in this work.

Nicholas Burgess
July 1981

ARTHROPODS OF MEDICAL IMPORTANCE

Introduction

In many parts of the world a group of the smaller arthropod animals, the mites, ticks, bugs, lice, fleas and flies, has always been and still is a most serious factor in the dissemination of disease. This is particularly so in the tropics, where in some regions these pests constitute a threat to the existence of mankind.

The attack may be physical, as when countless numbers of blood-sucking mosquitoes in the Arctic infest man during the short summer period. The biting rate of these mosquitoes has been recorded as 280 per minute.

Mainly, however, the menace of these pests arises from their complex associations with pathogenic viruses, fungi, bacteria, protozoa and helminths, which result, at some stage, in the organisms being introduced into the human body. This gives rise to diseases which may assume epidemic proportions.

Usually the arthropods themselves are not injured by harbouring the pathogenic organisms, probably owing to a long evolutionary association with them. In certain cases, however, the association is more recent, as with body lice which have ingested the rickettsiae of epidemic typhus from a human case of the disease, when the arthropod is as unable as man to accommodate the pathogens, and is finally killed by them.

Arthropods may convey disease to man in a number of different ways, ranging from the simple mechanical transportation of pathogenic organisms on the arthropod body, as when house flies carry dysentery bacilli from infected faeces to food, to the entirely different and complicated process of incubating the pathogens through part of their life cycle, within the body of the arthropod, before the latter is able to infect man. Such a process occurs when a female malaria-carrying mosquito ingests malarial gametocytes

9

from the blood of a person suffering from the disease, and after incubating them through the sexual stage, finally injects them into the blood of a fresh human victim in the form of malarial sporozoites.

Although scientific classification has been used as little as possible in order to avoid the use of unfamiliar terms, the text is carefully arranged so as to begin with the lowest forms of the arthropods of medical importance (mites and ticks), and as systematically as possible, to work through to the highest forms (flies).

However, as some insects and arachnids have no popular name, being known only by latinized words, it is necessary to state a few usages of scientific nomenclature.

Every creature belongs to a 'species' and is scientifically described by two consecutive words. For instance, *Culex pipiens* is the name of a species otherwise known as the common brown house-mosquito. The name means the piping or whining gnat (*pipo*, to pipe).

A specific name is invariably printed in italics, but when written in ink or pencil is underlined. The first word of the two is always begun with a capital letter, while the second word begins with a small letter.

The first word of the specific name, when separated from its fellow is known as the 'genus'. Most genera include several species. For instance, the genus *Culex* includes *Culex pipiens*, *Culex fatigans*, *Culex tarsalis*, etc.

Several related genera may for convenience be gathered into a 'tribe', which will always end in the letters 'ini' (*e.g.*, Culicini or Culicine mosquitoes). Related tribes will be included in a sub-family, which will end in 'inae' (*e.g.*, Culicinae, or mosquitoes).

Related subfamilies are always grouped into a family, which ends in 'idae', as for instance in the Culicidae or family of non-biting and biting gnats.

Related families may, if numerous or if it is otherwise considered necessary, be placed in a 'superfamily', which ends in 'oidea'. For instance Sarcoptoidea, the superfamily including scabies-mites and their relatives. Families and superfamilies are always grouped together in an 'order', which has no special ending, though the commonest termination is *'ptera'* (wings). An example is Diptera, the order of two-winged flies (*dis*, twice; *pteron*, wing).

Related orders are always included in a 'class' (which has no special ending), such as the Insecta, or class of creatures cut into a distinct head, thorax, and abdomen (*in*, into; *seco*, to cut).

Related classes are placed in a 'phylum' (which has no special ending), such as the Arthropoda or joint-footed creatures, (*arthron*, joint; *poda*, feet).

Finally all related phyla are grouped under one Kingdom, such as the Kingdom Animalia or breathing creatures (*anima*, breath).

The basis of this sytem was worked out in 1758 by Linnaeus, the great Swedish naturalist, and has never been bettered by any rival nomenclature.

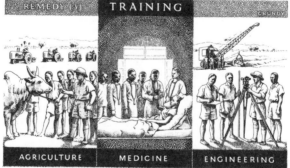

A series of four drawings made for the centenary of the Ross Institute in 1957 to show the role played by Britain in the development of some African countries.

CHAPTER 1

THE EVOLUTION AND IMPACT OF MEDICAL ENTOMOLOGY

Insects and arachnids attack man in two ways. Firstly, by Physical Attack, as when one or more of them bite or sting, and secondly, by Transmitting Pathogenic Organisms, as when a female *Anopheles* mosquito inoculates the parasites that cause malaria.

An example of 'physical attack' occurs each year during the short Arctic Summer of one to three months in northern Canada. Here, over an area of thousands of square miles, the thaw of the upper few feet of frozen tundra liberates vast quantities of water which gather into puddles, streams and lakes. In this shallow water mosquito larvae of the genus *Aëdes* (which specialises in breeding under adverse conditions), hatch from eggs which have remained dormant during the previous nine to eleven months of winter. By feeding voraciously on the plentiful supply of organic debris available, the aquatic larvae quickly pass through their four stages, then pupate, and finally become winged adult male and female mosquitoes. These mosquitoes are uncountable, for clouds of them may be seen as fog banks stretching from horizon to horizon, and as these Arctic *Aëdes* are among the largest and fiercest of their kind, it is easy to understand that no human work could proceed without the protection of special clothing and anti-mosquito measures. It is difficult, however, to appreciate the high rate of biting recorded of these mosquitoes, which rate is obtained by baring the hand and forearm to the elbow and with the aid of a stop-watch, counting the number of mosquitoes which alight to feed during a period of 60 seconds. By this method a rate of 280 per minute has been recorded. The ferocity of such an attack becomes apparent if the experiment is made of outlining an arm and hand on a blackboard or sheet of paper; then with a watch in hand, a series of chalk or pencil marks is dabbed rapidly between the outlines, in imitation of mosquitoes alighting to feed. By counting aloud and dabbing as quickly as possible a rate of approximately 180 may be achieved. When one considers, therefore, that not merely for one minute, but throughout the 24 hours of the day, clouds of large and hungry mosquitoes are endeavouring to pierce the defences of special clothing and repellents, it can be understood that this kind of physical attack seriously affects the operations of all human beings in these areas.

The late Professor Buxton once told me that when the permanent settlers of northern Canada became men with money in their pockets after working on the construction of army camps, they did not rush to spend their new wealth on provisions and luxuries, but on moving their homes to exposed promontories on lake shores, where the winds, uninterrupted by forest or scrub, were too strong for either mosquitoes or *Simulium* blackflies to remain for biting purposes. Here for the first time in their lives the children were able to play outside the cabins, a privilege previously impossible during the mosquito and blackfly seasons.

Simulium or biting blackflies have always been a seasonal menace in the tundra and forest regions of northern Canada, where their bites may cause an allergic condition known as 'bung eye', a swelling of the eyelids resulting in the victim being unable to see. This has been the cause of a number of deaths, when linked with exposure and shock through failure to make camp.

When the Danube basin suffers a drought year, conditions suitable for the mass breeding of blackflies may occur, and huge swarms of these small biting insects may drift on the prevailing winds to areas adjacent to the breeding grounds, killing by the shock of their multiple bites all animals unable to escape by going underground or into shelters. Men during 'fly years' protect themselves in the open by special clothing and by smokey fires termed 'smudges', or else by retreating into buildings. *Simulium* flies bite only by day, whereas mosquitoes typically bite during the night as well as by day if conditions are suitable.

Placing the problems of the physical attack of insects and their allies to one side, and examining instead the Transmission of Pathogenic Organisms by insect vectors, an entirely new world and concept of thought is encountered which has completely changed the course of most human societies.

For practical purposes, awareness of this world of microscopic organisms began with the Dutchman Leeuwenhoek, whose microscopes made it possible for scientifically minded men to observe structures previously invisible to the sharpest eyesight.

With the aid of his home-ground single (that is, not compound) lenses, he was able, among a host of other observations, to describe the red corpuscles of human blood, and to note that they were round in outline and not oval as in frogs. He further showed that weevil beetles in granaries and fleas in houses did not generate spontaneously in flour-dust or dirt respectively, but were developed from eggs, through grub stages which ate voraciously, and finally emerged as adults from a non-feeding pupa.

He also, in 1676, for the first time described 'animalcules' in drops of rain water, that is, bacteria and protozoa, and claimed that these were universal, being carried by particles of dust blown about by the winds.

In spite of the publicity given to these microscopic animalcules and the permanent interest they aroused, two hundred years were to pass before it was realised that some of them could cause specific diseases.

Later, by means of special culture and staining techniques, Koch identified the tubercle bacillus in 1882, and the cholera vibrio in 1884 with its typical transmission to man in drinking water.

The secret behind the discoveries of these and other important bacteria was simply the knowledge of how to stain them with aniline dyes, so that from a state of invisibility they could be seen under any reasonably high power microscope. To emphasise the importance of this usually overlooked discovery there is a record that Lister, after pondering Pasteur's work on putrefaction, entered a ward at three in the morning and lifted a patient's dressing to obtain a sample of pus. This he placed under his brass microscope, after removing the covering bell-glass, and gazed hopefully at the material on the slide. Of course, he was unable to distinguish anything, because he had not known how to stain the causative micro-organisms and so render them visible!

The techniques of staining were acquired slowly by medical men, perhaps because they needed demonstrations by an expert before understanding them, and as workers of different nationalities tended to be geographically separated, it was some time before the advancements of differential staining became common knowledge.

Thus it was, that although William Perkin, at the age of 18 years, had discovered mauve purple in 1856 whilst trying to synthesize quinine, and so found the first of the aniline dyes, it was not until 1879 that the German Ehrlich, first used acid and basic dyes (acid fuchsin and methylene blue) in combination for staining blood. Then in 1891, the Russian Dmitri Romanowsky used eosin and methylene blue for staining blood parasites, creating a standard that is still accepted today.

In spite of the advantage of this line of chemical research, most workers continued to study their microscopic preparations without differential staining. So in 1876 (the year Koch proved the anthrax bacillus to be the causative organism of the anthrax disease) Patrick Manson, first on Formosa and later on Amoy on the Chinese mainland, proceeded to study filariasis with its dramatic symptom elephantiasis, which he was able to do without the benefit of stains, since most parasitic worms are large enough to be seen in unstained preparations.

In the course of his investigation, he became more and more impressed with the 'microfilariae' in the blood of his filariasis patients exhibiting a nocturnal periodicity, as the immature worms were found to be swarming in the peripheral blood at night-time, though absent from this position by day.

Manson reasoned that this must be an effort of the parasites to obtain passage from man to man, and he could only interpret their coming to the surface at night as an effort to enlist some nocturnal blood sucking insect to act as carrier. This he decided must be a mosquito.

To prove this theory Manson fed *Culex fatigans* mosquitoes on patients with microfilariae in their peripheral blood. He then dissected these mosquitoes, and traced the development of the immature worms from inside the stomach of the flies, through their gut wall into their blood circulation, and finally into the insects' thoracic muscles, and observed that in this situation the worms grew larger, developed a mouth, an alimentary tract and other organs, and concluded that 'they were manifestly on the road to a new human host'.

Unfortunately, although this work was spread over seven years,

16

since his revised observations appeared in the *Linnean Society's Transactions* in 1884, Manson did not trace the passage of the mature microfilariae into the proboscis of the mosquito, from whence they pass to a new human host during the mosquito's next blood meal.

Being misled by a book on natural history which informed him that mosquitoes were as ephemeral as mayflies, he concluded they did not bite after oviposition but fluttered down to die on the water in which they had laid their eggs, and that filariasis was acquired by man through drinking water infected by such larvae as escaped from the bodies of the drowned mosquitoes.

Although an anonymous reviewer in the *Veterinarian* of March 1883 had suggested that if a mosquito could pierce human skin and suck up immature microfilarae it could possibly deposit the mature larvae on a new human host the next time it fed, Manson did not follow this lead, and it was left to the Australian Bancroft in 1889 to discover the true transmission cycle. Whilst Manson was still polishing his work on filariasis a historical discovery was made in regard to the most important disease in the world.

On 6 November 1880, the young French Army surgeon Alphonse Laveran, whilst stationed in Algeria, correctly identified the parasites that cause malaria, and in a preparation of fresh unstained blood, among other forms, recorded the 'act of exflagellation' as part of the life of a living parasite. This discovery greatly interested both Manson and young Surgeon Major Ronald Ross of the Indian Medical Service. Manson was particularly interested because the flagella was reminiscent of microfilariae, and Ross because malaria killed a yearly average of one and half million people in his India.

Although Manson did not himself hear about the parasites of Laveran until 1885, or see them until 1892 at the Seamen's Hospital at Greenwich (for a demonstration of method was usually needed to see these parasites for the first time in unstained preparations), yet he was in a position to show them to Ross in 1894, when Ross who had been unable to find them on his own account during a 12 year study of malaria, obtained an introduction to Manson, in order to be given a demonstration in his turn.

Manson was able to show the method of making a blood slide and recognising the parasites, at Charing Cross Hospital, and in

November 1894, as they were walking along Oxford Street, Manson turned to Ross and said 'Do you know I have formed the theory that mosquitoes carry malaria just as they carry filariae,' and then proceeded to explain that he considered the malarial flagella to be spores which enter water on the death of the egg laying female mosquito and so infect man when he uses the water for drinking purposes.

Ross was much impressed, and determined to investigate the matter on his return to India the following year. With the assets of enthusiasm, perseverance and a new microscope, but with the drawbacks of not knowing how to dissect mosquitoes, or to classify them into their different kinds, or how to stain microscopic preparations, Ross set to work in 1895, and in two years reached his first goal.

To begin with, he confused himself by experimenting with drinking water in which mosquitoes that had been fed on malaria patients had died. His first volunteer to drink such water contracted malaria, but the following 21 volunteers remained healthy! Ross's next attempts were to feed *Culex* and *Aëdes* mosquitoes on malaria patients, and to see if he could obtain the transmission of the parasites, either *via* the defaecations when deposited on human skin, or *via* the skin puncture while the female mosquitoes were feeding. These attempts failed, largely because Ross had used the wrong kind of mosquito, for it was not until later that he was to find that only members of the genus *Anopheles* are capable of harbouring the parasites of malaria.

Finally, however, on 20 August (known for years afterwards as 'Mosquito Day'), and again on 21 August 1897, he was at last rewarded by seeing rounded pigmented oöcysts on the stomach wall of some *Anopheles stephensi* which he had previously fed on a patient whose blood contained the crescent shaped female cells or macrogametocytes of falciparum malaria. These oöcysts he recognised immediately as malaria parasites because they were similar to the pigmented cells Laveran had described in 1880.

That he still accepted Manson's theory that the flagellate bodies of the male cells were spores, and that these were transmitted to man in drinking water does not detract from the fact that his first goal was the most important discovery mankind had yet been given, *viz* that an insect was at last proved to be the intermediate host of an organism responsible for a great epidemic disease.

The second goal was now obvious; to find how the parasites in the mosquito were conveyed to man. After some difficulty, Ross secured a post at Calcutta in 1898, but when he tried to get volunteers and patients for his experiments he could not obtain any cooperation, for the local population had recently been rioting against Mr Haffkine's anti-plague inoculations.

Ross, therefore, returned to bird malaria in sparrows. That he should do this so readily may have had some connection with an episode that occurred early in 1897. On this occasion he had witnessed a flagellum from a male parasite struggling inside a female parasite, but with Manson's spore theory in his mind he had dismissed the incident as that of a spore endeavouring to get out of a cell, instead of in!

The true sex significance of this was discovered the same year by a young American student William MacCallum who was studying avian malaria in crows. When Ross heard the truth of the act of exflagellation he wrote that 'he had felt disgraced as a scientist ever since!' With this in his mind it was natural that he should turn to bird malaria in the absence of human volunteers. Also being now freed from preconceived ideas of spores and drinking water, he was more able to trace the complete cycle—from man to mosquito, and from mosquito to man.

His careful investigations of the mature oöcysts on the mosquito stomach revealed that they finally burst *in situ* and evacuated their contents into the blood stream of the mosquito, so Ross set himself to find where these contents went. They appeared to consist of masses of tiny motile rods which had their greatest concentration in the thorax. Close observation, for Ross did not then know how to dissect a mosquito, finally revealed the rod-like organisms or sporozoites thickly packed in the cells of the salivary glands. Once this was seen it was obvious that parasite-laden saliva of the feeding mosquito would be infective to a susceptible host, and this was proved by Ross when he infected 22 out of 28 sparrows by inducing infected mosquitoes to feed on the birds.

His second goal was achieved in 1898, and the final goal of proving the mosquito transmission of malaria in man was achieved in the following year by Manson, who obtained some female *Anopheles* mosquitoes which had been fed on a malaria patient in Rome. These infected mosquitoes were allowed to bit two healthy volunteers (one being Manson's only son), in London, where there was no malaria, and as both developed the disease, the final proof of insect-borne transmission was beyond doubt.

Between the two dates 20 August 1897 and 13 September 1900 (when Thorburn Manson began his malarial fever in London) lies the dividing line between the Past and the Present in History, between the Ancient World and the Modern World.

Before the double ruled line of these two dates lay the long history of *Homo sapiens* during which he struggled blindly against fate and disease, when women had to breed continuously, as do the lower animals, in order that sufficient progeny should survive to assure continuance of the race. During this long period religions were largely rituals of supplication to cosmic gods to grant succour to those who prayed, and invariably these religions were accompanied by a symbol of terror called the Devil, named by the ancient Hebrews better than they knew Beelzebub 'King of the flies'. Before the discoveries of Ronald Ross, mosquito-borne malaria, flea-borne plague, louse-borne typhus, louse and tick-borne relapsing fevers, mosquito-borne yellow fever, fly-borne typhoid fever, the fly-borne dysenteries and fly-borne cholera, not only took their steady toll of life each year and from each generation, but every ten years or more appeared as epidemics, and every few hundred years in a world-wide pandemic form, as in the Black Death of the 14th Century and the Plague of the 19th Century.

After the year 1900 mankind has been able, not only progressively to discover the causes of the great epidemic diseases that before had seemed to be curses upon the human race, but also increasingly to find means for their control and eradication.

From this date also became possible our vast cities with their swarming populations, the huge armies of over ten million men in the field, the incredible revolutions of political and social life that are unique in the world's history, and our present achievements regarding the atom and the conquests of outer space. To some extent insects have now ceased to be man's greatest enemy, for he has assumed that role himself, since whatever he now fears most, he is himself responsible for creating.

20

Once insects and their allies were proved the carriers of epidemic diseases, either by simple mechanical or complicated cyclic transmission, it only became a matter of expending sufficient mental energy, time and money to find the causes and in many cases the cures for those arthropod-borne diseases which had previously kept mankind in such merciless subservience.

The foregoing is proof of the importance of the quite recently evolved science of Medical Entomology, without which most people alive today would not have been born, or if brought into the world would have been cut off in early life by epidemic disease. To explain this more clearly there are two aspects of medical entomology which it is vital to understand. First, the practical or functional aspect, where the entomologist tracks down and identifies the arthropod pest, finds out its habits and life history in relation to disease and endeavours to control or eradicate it. Second, the inevitable and usually unforeseeable consequences of this practical work.

A good example of the practical side of medical entomology is concerned with the invasion of tropical South America by the most infamous mosquito vector of tropical Africa *Anopheles gambiae*. This mosquito has two notorious characteristics. It thrives best in the vicinity of human habitations and it quickly colonises itself in shallow, sunlit rain-puddles. It is thus able rapidly to increase its range during and at the end of the rainy season, in a manner not known of any other mosquito.

Therefore, in March 1930, when Shannon the entomologist found some 2,000 *Anopheles gambiae* larvae in a pool in a hayfield behind a sea wall in north east Brazil, his report to the health authorities at nearby Natal city, that a dangerous vector had been carried from Africa to South America, seemed to him most urgent news. No particular notice was taken of this report however! Five weeks later, when the worried entomologist returned to examine the mosquito situation, he found that malaria had broken out in epidemic form, so that not only was it necessary to distribute medicines, but also food to enable the local population to survive. His survey at this time revealed that the vector mosquitoes had extended their range by about one kilometre, and the epidemic itself at last stirred the city authorities into action, so that the Brazilian anti-*Aëdes aegypti* yellow fever personnel were appealed to to help apply Paris green arsenical dust to the known *Anopheles gambiae* breeding waters. This plus the advent of the dry season, brought about a substantial lowering in the number of cases of malaria. The entomologist's recommendation that the sea walls be broken through to allow water flooding of the hayfields was resisted so persistently by vested interests that nothing was ever done in this direction, although this simple control measure would have quickly destroyed all the *gambiae* at their original focus. It is a historical fact that the vested interests of rich and poor always resist control measures that affect their income, no matter how serious the danger is! The entomologists now reported that *Anopheles gambiae* larvae were to be found over an area of over some six square kilometres, but little was done to limit their breeding until a devastating outbreak of malaria in January 1931 forced the Federal Government itself to take action, and with the aid of the Rockefeller Foundation a campaign was finally launched that eradicated *gambiae* from this area by 1932. Application of Paris green larvicide formed the basis of control in this instance.

Further surveys by entomologists failed to find any *gambiae* whatever, either adults or larvae, but mosquitoes are difficult creatures to overcome, and so instead of this being the end of the story it proved to be but a beginning, for soon after the 'all clear', rumours of epidemic outbreaks of rural malaria began to come in from the surrounding countryside.

Fortunately a barrier of open sea lies to the east of this north east tip of Brazil, while to the south and west are swift rivers, contrary winds and mountain barriers. At the north west, however, *Anopheles gambiae* somehow managed to slip through and colonize certain valleys which were much more suitable to its habits than its original point. However, as five years of drought concurred with this silent invasion, which drought was severe enough fully to occupy the Government in relief schemes for alleviating the suffering of the agricultural villages, the *gambiae* population, with its attendant malaria, was allowed to occupy more and more territory without any particular outcry.

Then the dry seasonal weather changed, and the flood rains began. The long inland river valleys became sheets of deep and shallow water. The roads were submerged and prevented ingress of supplies and egress of people wishing to escape, and under these conditions *Anopheles gambiae* became ubiquitous, and malaria exploded into full scale epidemics among the marooned population. The general belief was that the north east of Brazil would be completely depopulated by malaria, aided by starvation and emigration, and this tragic state of affairs lasted, in spite of national efforts to combat it, until the year 1939.

The devastation that occurred has been described as similar to Mauritius in 1866 and 1867, the Punjab in 1908, Ceylon in 1935 and but for prompt action in Egypt in 1945. This was simply the result of one species of *Anopheles* mosquito being allowed to establish itself in a new country despite the warnings of entomologists.

At last, however, the combined resources of the Brazilian government and the Rockefeller Foundation mounted eradication attacks with Paris green arsenicals which, starting at the perimeters of the stricken areas, gradually worked inwards, eventually destroying the invading *gambiae* by 1940.

It is noteworthy that at the time other South American governments were extremely loath to join any common fund for use against the *Anopheles gambiae*, either hoping that eradication would be successful before the invading mosquitoes reached their territory, or else out of carelessness, engendered by lack of imagination. This very common attitude is one of the main reasons why medical entomology should not be allowed to lapse when it is not actively fighting some outbreak, for like the fire service in a well organised community, one never knows when a situation will need controlling, and that quickly if there is not to be a catastrophe; for the official mind, fully occupied with administrative problems, is rarely in a position to organise quickly against disasters, if expert help is not at hand. That the complacent modern world is likely to run into insect-borne catastrophes in the near future is already threatened by the alarming rate with which arthropod vectors are developing genetic resistance to each new insecticide as it is introduced. DDT, BHC, Dieldrin and Lindane, are already ineffective in several parts of the world and chemists are hard pressed to keep ahead of these resistant strains.

As to the means by which *gambiae* reached South America, the evidence points to French destroyer postal vessels which made fast runs from Dakar in Africa, and regularly anchored off the particular portion of the coast where the *gambiae* were first found, and it is believed that these 'avisos' were responsible for carrying adult mosquitoes across the Atlantic. The break through of the vector to the north west, is believed to have occurred by coastal shipping or road vehicles carrying the adults, all of which shows how easy it is for man himself to be a 'carrier' of disease vectors!

Finally, after giving this example of practical entomology, there is left the most important aspect of all 'The inevitable and unforeseeable consequences of the practical work', for it must always be realised that medical entomology in practice alters the evolved balance of nature, as it has affected man, animals and plants during the past half a million years.

This shows itself in two simple and distinct ways. Firstly 'By allowing human populations to increase explosively, in the absence of the checks that diseases and starvation previously imposed'. Secondly 'By endowing the increased population with new mental health and energy' so that the people of countries who before were considered listless and peaceful, have become violently nationalistic or otherwise aggressively group-conscious and combative.

The first aspect, that of population increase, has been very seriously studied by persons and councils responsible for World Health Planning, and two clear schools of thought have emerged, both desirous of ultimately benefiting mankind. One school feels it is wrong suddenly to introduce the benefits of medical entomology and upset the age old balance in the densely populated areas of the world, until such time as these countries and continents have been taught how to limit their birth rate. For it is feared that these populations, suddenly freed from endemic and epidemic disease, may simply exchange death by disease for death by starvation.

The other school of thought feels strongly that help must always be given to rid populations of disease, and that such 'bridges' as arise in populations beyond food supplies are to be crossed when they are arrived at. Whichever method is adopted, the consequences to us personally and to our descendants is a matter for future life and death.

The second effect, that is increased mental health and energy has not received the same attention as the problem of world population, although its effects are much more topical. In this respect it is interesting to look at a map and note the world-wide stations of medical discoveries, many of them by army medical officers of different nationalities, and then to examine the recent political changes that have taken place in these areas.

The Dutchman Leeuwenhoek gave the microscope with which to see pathogenic organisms. The German medical officer Koch gave the cause of tuberculosis and cholera and its mode of transmission, after investigations in India and Egypt. The Scottish medical officer Manson discovered the mosquito host of filariasis and elephantiasis after work in China. The Australian Bancroft gave the complete transmission cycle of filariasis from Australia. The French Army surgeon Laveran recognised the parasites of malaria during work in Algeria. The Russian Romanowsky gave the differential stain by which blood parasites may be viewed and studied. The British army medical officer Bruce gave the tsetse fly carrier of nagana and sleeping sickness in Africa. The IMS surgeon Ross gave the mosquito transmission cycle of malaria, during studies in India, and so the list might go on with many names, including those who gave their lives directly in the cause of medical entomology.

These discoveries, however, as the 20th Century gathered way, became more and more associated with control measures, until we find in many countries that control or eradication schemes have replaced the endeavour to find new vectors and disease organisms, and the effects of this are worth thinking about.

To sum up, what does the impact of medical entomology ultimately mean to us and our children? It means that scientific mankind must learn to think in world wide terms or suffer terrible reverses of his own making. He must learn to consciously limit his numbers and he must continue to fight disease, not only because he considers human life to be sacred, but because a modern adult human being has cost so much in parental effort and social expenditure to bring to maturity. Furthermore, human beings who are living in this generation of machine miracles and scientific opportunities, must utilise their knowledge with a minimum of personal greed and lust for power, and with a maximum of gratitude and appreciation for those generous and dedicated students of science and the arts who have raised, and continue to raise, all mankind towards the appreciation of civilised culture.

CHAPTER 2

MITES OF MEDICAL IMPORTANCE

Order *ACARINA* (in part)

These are minute (excepting the Holothyridae and some adult *Trombicula*) eight-legged arthropods with unsegmented sac-like bodies which may or may not have a waist-like constriction. There are no winged forms; and unlike their larger relatives the ticks, they have no projecting hypostome below the mouth for anchoring in the host's skin.

'Forage Mites'

This is a popular name given to those species of *Glyciphagus* and *Tyroglyphus* (Fig. 2), including the common cheese and flour mites, which normally feed and breed on stored flour, oatmeal, other cereals, sugar, cheese, dried meats, etc., and which occasionally become parasitic on man.

Disease caused by Forage Mites. Occasionally a severe DERMATITIS accompanied by pruritis is caused by a mite invasion

Adult   Hypopus

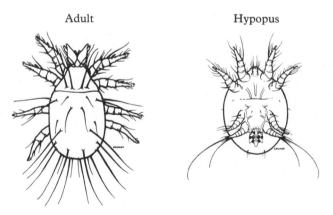

Figure 2   Forage mite, *Tyroglyphus*

of the skin in persons handling stored food products. 'Grocer's itch', 'Copra itch', 'Vanillisme', are three examples of local names for geographically separated outbreaks of mite dermatitis. Allergic hypersensitization is often built up during attacks, which may lead to severe recurrence of the condition when contact is made, even two or three years later, with the dead bodies or faeces of the mites in warehouse dust.

Intestinal disturbance is occasionally caused by ingestion of food containing a large number of mites. Forage mites have even been known to breed in the human alimentary tract.

Recognition of Forage Mites. They are just visible to the naked eye, of whitish colour, soft-bodied with fine bristles, distinctly longer than broad, and with the cuticle *not* striated transversely.

Habits of Forage Mites. They breed rapidly in vast numbers when the conditions of temperature, humidity, and decomposition of their food are suitable.

Life-history of Forage Mites. The females typically lay eggs on stored food products. The larva has six legs. The nymph has eight legs and may moult to become adult if conditions are suitable, or it may become a *hypopus*. This is a stage very resistant to adverse conditions. It has a hard shell, is without a mouth, has legs unsuitable for walking, and attaches itself by means of ventral suckers to any passing mouse or fly for free transportation to a better locality, where it will change into a further nymphal stage, feed, and then moult to become a sexually mature adult.

Spread of Forage Mites. This may be by the introduction of infested food into store-rooms, and by hypopus nymphs being carried by flies, etc.

Rhizoglyphus parasiticus. These mites caused 'coolie' or 'water itch' in workers on old and contaminated tea plantations in Assam. They are of the same family as forage mites.

Straw Itch Mites *(Pyemotes (Pediculoides)
ventricosus)*

This mite (Fig. 3) normally feeds on crop-pests or grubs found in grain, hay, etc. It sometimes causes a severe DERMATITIS and pruritis termed 'grain-itch' in workers discharging grain or cotton cargoes, and may be introduced into buildings in mattresses or furniture stuffed with hay or straw.

Unfed female   Gravid female

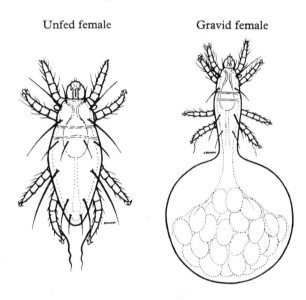

Figure 3   Straw itch mite, *Pyemotes ventricosus*

Recognition of Straw Itch Mites. The adult male is just visible to the naked eye. It is pale yellow in colour and of a somewhat flattened and angular appearance. The dorsal surface bears four pairs of bristles. The young female is longer and more cylindrical than the male, and *it is this stage that attacks human beings.* The gravid female, similarly pale yellow in colour, is easily visible to the naked eye, and in appearance resembles an insect's egg or grass-seed attached to the skin. The abdomen becomes enormously distended with the developing young until it is 20 to 100 times as large as the rest of the body.

Life-history of Straw Itch Mites. The eggs hatch in the uterus of the female and the young reach sexual maturity within the parent before they are born. There may be as many as 270 mites, born at the rate of 50 or so a day. The life cycle takes about six to ten days.

Modes of infestation. The mites may be acquired by handling infested grain, beans, cotton, etc., also from wind-blown dust during the unloading or harvesting of such material; also by contact with infested straw, or by contact with a human carrier who may transfer some of the active females.

Some Other Blood Sucking Mites

Dermanyssus gallinae. Mites of this species are often termed 'red poultry-mites'. They may become temporary parasites of man, and cause by their bites a transient DERMATITIS; and intense pruritis known as 'poultryman's itch'.

Disease caused by Red Poultry-Mites. They are natural transmitters and reservoirs of ST. LOUIS ENCEPHALITIS among chickens. Mosquitoes carry the virus from chicken to man. (U.S.A.).

Recognition of Red Poultry-Mites. These are the small active red mites found in birds' nests and cages, and in poultry runs. The females have the two processes of the visible mouthparts fused together to form a long fine style for piercing the skin of the host. The males have both processes separated.

Habits of Red Poultry-Mites. They are nocturnal and blood sucking and only breed freely on avian hosts.

Modes of Red Poultry-Mite infestation. Persons working in poultry houses or handling or plucking infested fowls are often attacked. The nearness of poultry or pigeon hosts to human sleeping quarters sometimes leads to invasion by the mites.

Liponyssus bursa. Mites of this species are sometimes named 'tropical fowl-mites' as they replace the common red mites in some warmer parts of the world. They cause a similar DERMATITIS and pruritis to those produced by red poultry-mites.

Liponyssus bacoti. This species of mite is sometimes known as the 'blood sucking rat-mite', on which rodent it is normally parasitic. These mites not infrequently attack man, inflicting painful bites which may give rise to a severe pruritis or 'RAT-MITE DERMATITIS'. This species is suspected of conveying ENDEMIC TYPHUS from rodent to rodent and possibly to man. It can also transmit RICKETTSIAL POX. Hereditary transmission through the eggs may occur in both these diseases.

Recognition of Rat-Mites. These mites are without any indication of a waist, and have the visible mouth-parts pincer-like in both sexes.

Modes of Rat-Mite infestation. The presence of numerous rats or mice is always a danger. Fumigation with hydrocyanic acid gas may succeed in killing the rodents without destroying the mites, which will later swarm and feed hungrily on the nearest available hosts.

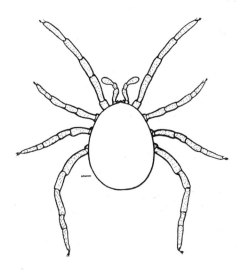

Figure 4   Red poison mite, *Holothyrus*

Holothyrus Mites

Mites belonging to this genus (Fig. 4) found in islands of the Madagascan, Oriental, and Polynesian regions, have caused several cases of POISONING. These have been chiefly among children who have touched the mucous membrane of their mouth after handling the mites.

Recognition of *Holothyrus*. These mites are exceptional in their size, since sometimes they reach $6\frac{1}{2}$ mm. in length. They are red in colour, perhaps as a warning, for they are poisonous to ducks and geese and possibly to other animals. Apparently the mites secrete an intensely irritating fluid which will remain potent on the fingers for some hours, during which time it may be accidently conveyed to the mouth.

Scabies Mites *(Scarcoptes scabiei* var. *hominis)*

Mites of this variety (Fig. 5) are sometimes termed 'human itch-mites' since they breed only in cutaneous burrows in the skin of man. Very similar mites are found infesting the skin of sheep, cattle, rabbits, cats, rats, horses, poultry, apes and other animals, where they cause 'mange'. These mange mites occasionally attack persons in contact with infested animals and cause erythematous patches with itching. As the mites do not breed in the human skin the condition does not last long.

Diseases caused by *Sarcoptes scabiei*. 'SCABIES' is caused partly by the adult female mites eating and scraping major burrows in the human skin and causing itching by their acid secretions and defaecations; and partly by the secondary burrowings of the males and young females. The chief causes of the irritation, vesiculation and pruritis, however, are believed to be due to an acquired sensitivity leading to secondary infection by scratching. Sensitivity to the parasites usually develops in about one month in persons previously uninfested. In persons who have suffered an attack in the past, the itching develops within 24 hours of a new infestation. The first signs are reddish, slightly elevated tracts on the skin (where this is thin), leading from minute orifices to the whitish oval females, which may be just visible (by hand lens) below the skin at the blind end of the burrows. The lesions develop very rapidly as the breeding rate is high, so that there soon appear multiple papules, vesiculations, pustules, excoriations, and secondary infection through scratching. There are also generalized areas of sensitization which occur apart from the sites of active parasitization. The fertilized females make their major burrows in the human skin chiefly where it is thin and folded, *i.e*, between the fingers and toes, on the backs of the hands, the front of the wrist (approximately 60% are found on the hands and wrists), the elbows, the front of the axillae, the waist, beneath the female breasts, in the groin, on the scrotum, the shoulder blades, the small of the back, the front of the knee, and so on. Frequently, however, areas of thick cuticle are attacked. In children the lesions may appear anywhere. The major burrows are made during the warm period when the host is in bed. This may cause serious loss of sleep. Extension of the tortuous burrows is at the rate of about 2 or 3 mm. a day, until a total distance of some 15 mm. may be covered.

Male   Female

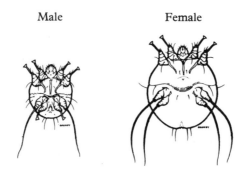

Figure 5   Scabies mite *(Sarcoptes scabiei)*

Recognition of *Sarcoptes scabiei*. These mites are just visible to the naked eye, the male mite being smaller than the ovigerous female. They are yellowish-white in colour and almost spherical in shape. The dorsal surface of the cuticle is transversely striated and bears a number of quite short spines. The anterior and posterior pairs of legs are widely separated and are very stumpy in appearance. The anterior two pairs of legs end in delicate unsegmented pedicels with terminal suckers. The posterior two pairs of legs in the female end in long bristles, while in the male the third pair ends in bristles and the fourth in pedicels and suckers.

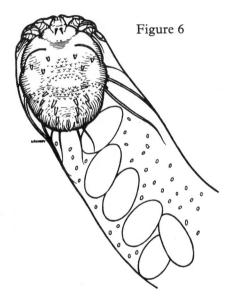

Figure 6

Adult female *Sarcoptes scabiei* mite in burrow in host's skin—note eggs and faeces

Habits of *Sarcoptes scabiei*. It is believed that the male mite finds and fertilizes the female on the surface of the host's skin. Young females occasionally wander over the skin. Gravid females are capable of ovipositing away from the host, and the eggs of developing in fomites.

Life history of *Sarcoptes scabiei*. The gravid female develops one or two eggs at a time in her body. She lays about 40 rather large, ovoid, transparent eggs behind her in her burrows during her month or more of life (Fig. 6). The eggs hatch in three to five days. The larvae are six-legged and moult after two or three days. The nymphs are eight-legged and there may be two nymphal stages.

33

Both larvae and nymphs are commonly found in hair follicles. The life cycle takes from eight to fifteen days, from the laying of the eggs to the major burrowing of the fertilized female. The average length of life in both sexes is four or five weeks, but it may be as long as seven weeks. In soiled undergarments the mites may survive for ten days.

Mode of infestation with *Sarcoptes scabiei.* The eggs, larvae, nymphs and adult mites may be acquired by contact with infested persons or their bed clothing. The newly fertilized females are chiefly responsible for the spread of scabies, as they become active and move about at night under warm conditions. Prolonged and close contact of body parts with an infested person under warm conditions are optimum for the transfer of Scabies mites. The fertilized female takes about one hour to bury itself in the horny layer of the human skin.

Control of *Sarcoptes scabiei.* The use of chemicals such as benzyl benzoate is effective in killing mites. The disinfestation of the patient's bedding and clothing is of minor importance compared with treatment and is not practised as a routine. Mites are not likely to survive for long in the bedding or clothing of treated cases.

Demodex folliculorum. This is a minute species, easily recognized on account of its highly modified shape, stumpy legs, and the transverse striations of the cuticle. It is found in blackheads and sebaceous glands. Though rarely causing a dermatitis in man, it is well known as the cause of dog and cat follicular or demodic mange.

Harvest Mites *(Trombicula)*

Mites of this genus are variously known as 'harvest mites', 'harvest bugs', 'red bugs', 'chigger mites', 'kedani mites', or 'bête rouge'. They are widely distributed in temperate, sub-tropical and tropical zones.

Disease caused by *Trombicula.* The larvae cause an intense itching often accompanied by severe DERMATITIS and occasionally by fever, when they attack man. Scratching may lead to secondary infection.

Some species, of which the most important are *Trombicula akamushi* (Japan) and *Trombicula deliensis* (Burma, Malaya etc.) in the Oriental, Australian and Polynesian regions, transmit to man *Rickettsia orientalis* the causative organism of MITE BORNE TYPHUS, sometimes referred to as scrub typhus, Japanese river fever, or tsutsugamushi fever.

These mite larvae are particularly prevalent in secondary jungle, *i.e.*, in land once cultivated from virgin jungle but allowed to revert. Such places often harbour numbers of rats, which serve as hosts for the *Trombicula* larvae.

Recognition of Adult *Trombicula*. The adults are mostly small but quite easily seen with the naked eye. They are reddish, hairy or velvety mites, with a noticeable constriction or waist. The terminal segment of the first pair of legs is always longer and more swollen than the other leg segments.

Habits of Adult *Trombicula*. The adult mites are harmless to man as they do not suck blood. Similarly the nymphal forms do not attack man. Both nymphs and adults are capable, however, of carrying an hereditary strain of typhus organisms and passing it through the egg stage to the next generation of biting larvae.

Figure 7   *Trombicula* mite larva

Recognition of *Trombicula* **Larvae.** These are so minute as to be invisible to the naked eye (Fig. 7). Clusters of larvae in the ears of rodents appear as orange or yellowish patches. When examined

35

MITE-BORNE TYPHUS *(Riekettsia orientalis)*
LIFE CYCLE OF TROMBICULA MITE

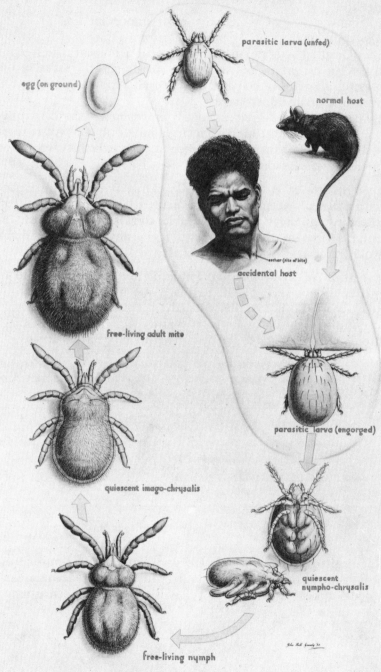

egg (on ground)

parasitic larva (unfed)

normal host

accidental host

eschar (site of bite)

free-living adult mite

parasitic larva (engorged)

quiescent imago-chrysalis

quiescent nympho-chrysalis

free-living nymph

under a suitable microscope the larva is seen to have no waist constriction. There are six legs terminating in trifurcate claws. Somewhat forward on the dorsal surface is a rectangular shield bearing seven pinnate hairs or processes. The dorsal surface posterior to the shield carries more than 30 setae or hairs.

Habits of *Trombicula* **Larvae.** These little creatures occur individually in vast numbers on vegetation or soil in their chosen habitat. They commonly attack persons gathering wild black-berries. The larvae tend to crawl up the legs until a clothing obstruction is reached. Here they attach themselves to the skin by their mouth-parts, and with the aid of the lesion remain at right angles to the skin for several days until engorged.

Life-history of *Trombicula.* The life histories of these mites are not well known. The eggs are usually laid on the ground. The larvae are six-legged and soon seek an arthropod, reptile, bird, rodent or perhaps human host and attach themselves for a pro-longed meal lasting several days. After this the larvae drop off and moult to become eight-legged nymphs and after a further moult become adults. There is one generation in a year.

Control of *Trombicula.* The principal measure for the control of mite typhus is the use of synthetic contact repellents such as di-methyl phthalate and di-butyl phthalate. This liquid, when lightly smeared over the clothing, immobilizes larval mites in a few minutes and kills them within half an hour so that the risk of their being able to bite is small.

Di-methyl phthalate is even more effective in killing mites but does not persist so well. Clothing impregnated with mite-repellent is effective for a fortnight and is not affected by washing two or three times in cold or warm water. Both di-methyl and di-butyl phthalate have a solvent action on plastics and must therefore be kept away from watch-glasses, straps and fountain pens made of this material. Benzyl benzoate is also an effective repellent.

Other measures are the wearing of protective clothing, *i.e.*, long sleeves, slacks, boots and gaiters or short puttees; the avoidance, if possible, in forward areas of places which are likely to be infested; the clearance and occupation of rear areas. These will in themselves greatly diminish the incidence of infection.

CHAPTER 3

TICKS OF MEDICAL IMPORTANCE

Order *ACARINA* (in part).
Superfamily Ixodoidea

These are large mites with a noticeably leathery integument. They have a visible spiracle on either side of the sac-like body in the vicinity of the fourth pair of legs. The false head (capitulum) never has eyes but bears a projecting lower lip (hypostome) roughened with sharp teeth for anchoring the tick in the skin of the host. There are always four pairs of legs in the adult forms. All ticks live by sucking the blood of amphibians, reptiles, birds or mammals. Except when fully gorged they are distinctly flattened from above downward.

Soft Ticks (Family Argasidae)

These ticks are found locally in many parts of the world, but primarily in sub-tropical and tropical regions. Some species multiply under very dry conditions.

Diseases caused and transmitted by Soft Ticks. The bite of some species of soft ticks may be painless, whereas that of others may be extremely painful and venomous. *Ornithodorus coriaceus* of California and Mexico is more dreaded than a poisonous snake because of the pain and lasting ill effects of its bite. Some soft ticks, such as *Argas persicus*, the Persian fowl-tick, have considerable nuisance value because of their readiness to bite man, particularly at night. One species, *Otobius megnini* the spinose ear-tick, occasionally invades the ear of man and causes severe pain. Hæmorrhages from the punctures, due to anti-coagulin, and ulcerations from injected toxins, variously result from the bites of different species of ticks.

Tick-borne relapsing fever. This is an acute recurrent fever caused by spirochaetes. It is widely distributed over the world and is transmitted to man by many species of the genus *Ornithodorus*.

Ornithodorus moubata is the most important vector, owing to its close association with the habitations of man in Africa. The disease is acquired through the bite of an infected tick of any stage, and also by contact with *coxal fluid*, which is an excretion from openings on the underside of many soft ticks and which is voided during, or soon after the blood meal. Once ticks become infected they remain so for the remainder of their lives and spirochaetes are passed to the succeeding generation through the sperms and the egg stage. Relapsing fever is thus endemic in ticks (though louse-borne relapsing fever is not endemic in lice, since the organisms eventually kill the louse). Tick-borne relapsing fever is not known to be infectious to lice. The animal reservoir of the disease is believed to include burrowing rodents, monkeys, squirrels, bats, and various cave-frequenting hosts of *Ornithodorus* including domestic animals such as goats, cattle and sheep if these rest in caves.

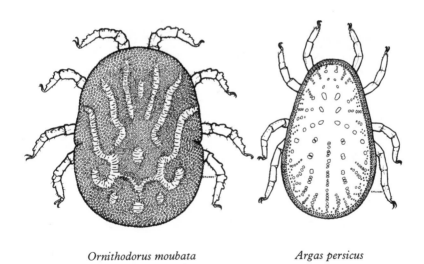

Ornithodorus moubata   *Argas persicus*

Figure 8   Soft Ticks, Argasidae

Recognition of Soft Ticks. Soft ticks (Fig. 8) are so named because they never have a hard chitinous shield (scutum) on the dorsal surface of their body. The mouth parts and false head of soft ticks are situated on the ventral surface and are not easily seen from above. These ticks often appear relatively shrivelled when starved but become plump and taut when engorged with blood. The adults vary greatly in size in the different species, ranging from three millimetres to almost thirty millimetres. The single spiracle on either side of the body is situated anterior to the base of the fourth pair of legs. The legs never develop a sticky pad (pulvillus) at their extremities. The two sexes in the soft ticks are very similar in appearance. The males are relatively smaller, and have a slightly different genital opening, but otherwise they resemble the females in outward appearance. Both sexes endeavour to engorge with blood at frequent intervals.

Habits of Soft Ticks. They are all blood-sucking, and usually 'multi-host' ticks, *i.e.*, they feed frequently on different individual animals, whenever these are available. They attack chiefly at night when the host is lying down. After feeding for several minutes or

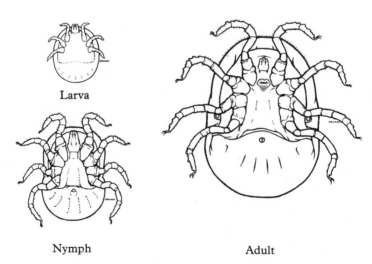

Larva

Nymph   Adult

Figure 9   *Ornithodorus* Soft Ticks (ventral views)

for a few hours, they retreat to the dust, sand, or crevices of the hut, cave, or other ground they have chosen to infest, and then bury or otherwise hide themselves. Here they digest their blood meal, moult, or lay eggs, according to the stage reached in their life history. A peculiar and important feature of some species of soft ticks is the copious secretion of coxal fluid which is passed either during or shortly after a blood meal. The clear fluid (in one species a crystalline powder) which appears to be the separated plasma of the host, issues from the opening of a pair of glands at the base of the second pair of legs. The tick's underside and the host's skin become wet with this fluid, which, in a tick infected with relapsing fever is laden with active spirochaetes, so that transmission easily takes place through the bite puncture.

Life-history of Soft Ticks. They are often referred to as 'ticks of the habitat', as they spend all their life, when not feeding, in the dust, sand, or crevices of their habitat. The globular, dark brown shiny eggs are laid on the ground or in sheltered places, away from the host, in several relatively (as compared with the hard ticks) small batches. The tiny larva (Fig. 9) is always six-legged and has no breathing spiracles. The mouth parts at this stage are more terminal and less ventral than in the adult. In *Ornithodorus moubata*, the important vector of African tick-borne relapsing fever, the larva remains within the open egg-shell and does not feed. In most species, however, this stage is very active. The nymph is always eight-legged, has one pair of breathing spiracles, and feeds frequently on available hosts, though never remaining on them for long. There are a number of nymphal stages with a corresponding number of moults (there is only one nymphal stage in hard ticks). The adults are eight-legged and have an opening to the sex organs situated between their leg bases; and therefore, near the false head (capitulum) and mouth parts. Copulation takes place away from the host. At intervals the females lay several comparatively small batches of eggs, which are invariably buried. There may be one batch after each blood meal, or several batches following one meal. The period of the life cycle varies in the different species and also according to temperature, and availability of blood meals.

Ornithodorus moubata, the 'eyeless tampan', may lay about 300 eggs in her first batch and about 30 less fertile eggs in her sixth batch. The eggs are laid at night and buried in the sand or dust. The larvae may hatch in approximately eight days, though they will remain in their egg-shell and not feed. The blood sucking nymphs will not appear for another four days or so. The period of time between each moult varies, as it depends on the availability of the blood meals preceding the moult. There may be four to seven nymphal stages, these requiring a few days rest after each moult before the ticks are willing to feed. The life cycle under good conditions is about 12 weeks. Most species of soft ticks can withstand prolonged starvation, in some cases for years.

Modes of infestation with Soft Ticks. Unfed soft ticks will readily attack any resting or sleeping person near their habitat, which may be deer-runs in forests, watering places for cattle and men, or caves and dwellings. They crawl or run to their victim, puncture the skin, engorge rapidly with blood (ten minutes to a few hours), typically excrete coxal fluid, and retreat.

Hard Ticks (Family Ixodidae)

These are found in all parts of the world where conditions are suitable. Usually a certain degree of humidity, not found on well-cultivated ground, is necessary for their multiplication.

Diseases caused and transmitted by Hard Ticks. The BITES of hard ticks often produce serious results, although the insertion of the somewhat large mouth parts is not usually painful. An engorged female hard tick may take up to a hundred times her original weight in blood. Bleeding from a tick bite is not easily stopped as the blood does not clot readily. If a hard tick is violently removed, the toothed mouth parts, together with the false-head, are often left in the host's flesh. This may result in secondary infection and ulceration.

'Tick-bite fever' is the term used to describe the general fever symptoms caused by the bites of ticks in susceptible persons.

'Tick paralysis' is a peculiar condition mostly found in children. It is due to the prolonged bite of some kinds of female hard ticks at a certain period of their development. The hollow at the back of

42

the neck is the most dangerous place of attachment. The symptoms are those of intoxication accompanied by an ascending motor paralysis. The patient dies when the organs of respiration become affected. *Dermacentor andersoni* and *Dermacentor variabilis* in North America, *Ixodes pilosus* in South Africa, and *Ixodes ricinus* in Europe are known to cause tick paralysis.

Tick-borne typhus (Spotted Fever) is a non-contagious, often fatal disease, transmitted to man by the bites of certain hard ticks. The causative rickettsiae can be passed through the egg stage to succeeding generations of ticks. The mammalian reservoirs of the rickettsiae are believed to be chiefly wild rodents such as rabbits and squirrels. Most of the large mammals are immune.

Q fever is a rickettsial disease that occurs in some parts of the world, for example, Australia, North America, Africa, Panama, Italy, the Balkans, Greece, Spain, Germany. It is transmitted to man mainly in the faeces of infected ticks. This material may enter bite-punctures or abrasions. The most important source of infection, however, is dry air-borne tick faeces, which readily enter the respiratory tract. Milk and butter have also been found infected.

In Australia the hard tick *Haemaphysalis humerosa* keeps the infection going in bandicoots and other bush animals, from which reservoirs the hard tick *Ixodes holocyclus* transmits the rickettsiae to man. In North America, the hard ticks *Dermacentor andersoni*, *Dermacentor occidentalis*, and *Amblyomma americanum* have been found infected. *Ornithodorus moubata* (by faeces and coxal fluid) and *Ornithodorus hermsi* soft ticks are good experimental vectors, and *Otobius megnini* (nearly related to *Ornithodorus*) has been found naturally infected.

Tularaemia, like plague, is an infectious disease of rodents and is transmitted to man by certain species of hard ticks in the Nearctic, Mongolian Palaearctic, and Japanese Orient. It runs a long course and is sometimes fatal.

Several *Dermacentor* and one *Haemaphysalis* hard ticks are incriminated, but the disease may also be transmitted by gadflies, as well as being water-borne (infectious when handled or drunk). Beavers and water rats are reservoir hosts in Russia, and mosquitoes have a part in the dissemination complex.

43

Recognition of Hard Ticks. Hard ticks (Fig. 10) are so named because they always have a hard, chitinous shield (scutum) on the dorsal surface. The mouth parts and false-head are terminal and easily visible from above. They have also the large single spiracles on either side of the body situated *posteriorly* to the base of the fourth pair of legs. A small sticky pad (pulvillus) will be found at the extremity of each leg. The two sexes in hard ticks are markedly different. The males have the chitinous shield covering the whole of their dorsal body surface (they do not engorge with blood), while the females have the shield only on the anterior part of the dorsal body surface. The posterior portion remains soft so as to allow for distension with blood and developing eggs.

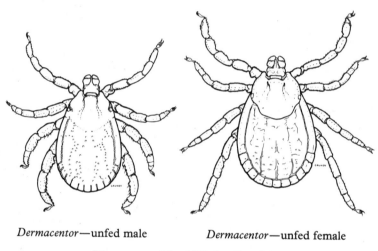

Dermacentor—unfed male   *Dermacentor*—unfed female

Figure 10   Hard Ticks, Ixodidae

Habits of Hard Ticks. They are all blood-sucking, and may be 'one-host' ticks, when they spend their larval, nymphal and adult stage on a single host, as does the cattle tick *Boophilus annulatus*, or they may be 'two-host' ticks, as is *Hyalomma aegyptium*. The majority of hard ticks, however, are 'three-host' ticks, of which the notorious *Dermacentor andersoni*, the 'Rocky Mountain spotted fever tick' is a good example. This species attaches itself when a larva or 'seed tick' (relative size and appearance of a seed), to a passing rodent and engorges itself for about a week. It then drops off and moults on the ground, producing the nymphal form which normally spends the winter without feeding. During the warmer part of the second year, the nymph endeavours to attach itself to another passing rodent, and if successful, engorges itself and drops to the ground. Here it moults again and becomes an adult, spending the second winter in an unfed condition. The third year, with good fortune, will see the adult *Dermacentor* attach to some large mammal such as a coyote, deer, bear, steer or human being, upon which host it hopes to find a mate, and if it is a female to engorge with blood, and after fertilization to swell with eggs to a relatively enormous size. Finally the gravid tick drops to the ground and seeks a suitably warm and humid place to deposit her eggs. Only one prolonged blood meal is taken at each of the three stages. Hard ticks at any stage are able to endure starvation for a considerable time, in some cases for years.

Life-history of Hard Ticks. They are often referred to as 'ticks of the host', as they spend a great part of their active life attached

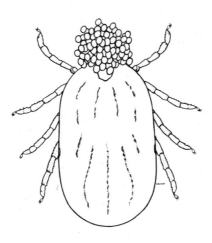

Gravid hard tick laying her egg mass

Figure 11

45

to the body of some animal. The globular light-brown shiny eggs are laid on the ground or in some sheltered place, away from the host, in one huge batch of several thousand eggs, over a period of several days (Fig. 11). The shrivelled female dies partly buried beneath her egg-mass. The tiny larva is six-legged and is without breathing spiracles. This stage is very active and often the whole brood attaches to one passing host. The nymph is always eight-legged and possesses breathing spiracles. Many nymphs fail to find a host, as some individuals will have dropped from their larval hosts in unsuitable places. The adults are eight-legged and have an external opening to the genital organs situated forward on the underside (venter) between the leg bases, and therefore near the false-head. The period of the life cycle varies according to the temperature (hard ticks hibernate in winter), and the availability of passing hosts. Adults of *Dermacentor andersoni* have been known to live for four years without food.

Modes of infestation with Hard Ticks. Hard ticks which have moulted on the ground and which require a further blood meal usually climb a grass stalk or stick and lie in wait. The first pair of legs of the tick—usually longer than the rest—become very active in the presence of a potential host, and will quickly seize hair or clothing fibres if the host comes within reach. They are chiefly active by day (soft ticks by night). Resting on ground or vegetation inhabited by ticks is a sure way to become infested. A light coloured woollen blanket dragged over such areas is the entomologist's method of capturing unfed wild ticks.

Control of Ticks. Residual insecticides are less effective against ticks than against insects, but some have given good results. For the control of soft ticks, the insecticide is applied to the habitat, *e.g.*, the floors of huts, as a dust or residual spray; for hard ticks the animal hosts, dogs and sheep, are treated with insecticides.

CHAPTER 4

BUGS OF MEDICAL IMPORTANCE

Order *HEMIPTERA*

Families Reduviidae and Cimicidae

These bugs may be winged or wingless, but they always have a proboscis or beak characteristically adapted for piercing and sucking. This is hingèd *below* the head when the bug is not feeding, and hinged *out* towards the host when a blood meal is being taken. All bugs have a distinct head, thorax, and abdomen, and three pairs of walking legs.

Cone Nose Bugs (Triatominae)

The distribution of the important species is in the Nearctic and Neotropic Americas; although allied forms occur in the Hawaiian, Polynesian and Australian Oceania; the Chinese, Malayan and Indian Orient; and the Madagascan, Tropical and South Ethiopian regions.

Diseases caused and transmitted by Cone-nose Bugs. The bites of these bugs are usually painless, but may cause urticaria if the host becomes sensitive to the injected saliva. Some species, however, habitually give painful bites with prolonged after effects.

American Trypanosomiasis or Chagas' disease, is caused by certain trypanosomes which are transmitted to man in the faeces of cone-nose bugs. The bugs habitually defaecate the remains of the previous meal on the host's skin while imbibing a fresh blood meal. This may lead to the organisms infecting man through the bite puncture, abrasions of the skin, or even the mucous membrane of the mouth, either by direct contact or through rubbing and scratching. Cone-nose bugs become infective some six to fifteen days after ingesting the trypanosomes, and may remain infective for as long as two years. The chief animal reservoirs of the disease in America

47

are armadillos and opossums as well as bats, cats, dogs, wood rats, and house mice. The most important vectors are *Panstrongylus megistus* (Brazil), *Triatoma infestans* (Brazil, Argentine, Chile, Peru), and *Rhodnius prolixus* (Brazil, Colombia, Mexico).

Recognition of Cone-nose Bugs. They are typical bugs, of medium to large size (averaging about an inch in length). They are usually of a chocolate brown colour. Some species have the abdomen below the wings barred with red or orange. The general appearance is of a longish, flattened, six-legged creature, which is usually seen in a motionless attitude or else making a short rapid run. The head is long and conical with a beak hinged from the anterior end. This piercing and sucking tube does not rest tightly against the underside of the head, but slightly away from it. The

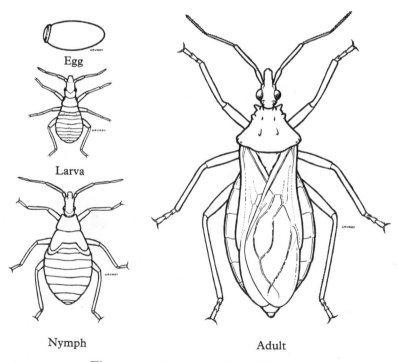

Egg

Larva

Nymph   Adult

Figure 12   Cone-nose bug, *Triatoma*

prothorax is prominent. In the adults the strongly veined fore-wings are folded tightly over one another and flat on the back. When at rest they completely cover the membranous second pair of wings. The nymphal forms have external wing buds only (no true flying wings).

Habits of Cone-nose Bugs. They all feed on the blood of animals. Some species are highly domestic and these tend to remain in the one house as long as it is inhabited. They remain in crevices by day and come out to feed at night. The adults are powerful fliers and able to alight for a blood meal on the faces of people sleeping above the ground, as in a hammock. They may similarly migrate from an uninhabited to an inhabited house. If the host (man or animal reservoir) has *Trypanosoma cruzi* in the peripheral blood, the organisms will be ingested by the feeding bug, and, following certain cyclic changes, be passed out in infective faecal deposits. The feeding time is short, ten minutes to half an hour. The incubation period of the organisms within the bug is six to fifteen days.

Life-history of Cone-nose Bugs. The fertilized females, about one month after mating, lay scattered eggs in cracks and holes in walls or other suitable hiding places (Fig. 12). The young larvae and the nymphs come forth to feed after dark and, being unable to fly, seek sleeping animal hosts by running and climbing. The different species have varying times for their life cycles. A typical life history is that of *Panstrongylus megistus* a highly domestic bug and well known in South America as a carrier of Chagas' disease. The eggs are quite easily visible, ovoid or barrel-shaped with an operculum or cap, and of a creamy white colour. They are laid at intervals in batches of about ten eggs. About 200 eggs are laid in all. The larvae hatch in a fortnight to a month, and feed about a week after hatching. They resemble the adults in miniature, but are without wings, or openings to the sex organs. There are five immature stages (including the larval), taking in all about 27 weeks. After the fifth moult the winged adults emerge and a week later they are ready to take a blood meal. Under favourable conditions the cycle from egg to adult is about ten months. The life cycle of the other species is known to range from four months to about a year.

Control of Reduvid Bugs. These bugs, which are the vectors of Chagas' disease in S. America, may be controlled by the use of residual insecticides in infested houses.

Bed-Bugs (Cimicidae)

These bugs may be found in houses, camps and ships all over the world. They are most abundant in temperate zones.

Disease caused by Bed-bugs. The BITES of bed-bugs are extremely painful and lasting in some persons, and occasionally result in transient dermatitis complicated with secondary infection. As the feeding time is usually at night, loss of sleep on the part of the human host is common. Neurasthenia has been recorded. Although suspected of transmitting many diseases, bed-bugs have not been proved vectors of any disease.

Recognition of Bed-bugs. They are about a quarter of an inch long, brown in colour and with an oval contour. After a blood meal they are darker and longer. A characteristic of bed-bugs is their much flattened appearance, especially when unfed. They are able to run rapidly when disturbed or when foraging for a blood meal. The piercing and sucking beak is hinged into a groove on the underside of the head when the bug is not feeding (in cone-nose bugs the beak does not fit into a groove). The compound eyes are black and prominent. The pronotum is markedly developed at the sides into thin shoulder-like cusps. Bed-bugs are wingless in all stages, though traces of wing cases are still visible, the power of flight having been lost because of their specialized mode of life. Nymphs and adults give off an odour from their stink-glands which characteristically taints a room harbouring large numbers of them.

Habits of Bed-bugs. Bed-bugs invariably hide by day in cracks and crevices of dwellings and issue forth at night to obtain blood meals. The feeding time of the adults is short, being about ten to fifteen minutes (less in nymphs). Bed-bugs rarely pass faeces on the skin while feeding. In unhygienically kept premises countless numbers may be found behind wall-paper, in cracks of plaster, in furniture or in other suitable nooks and crannies.

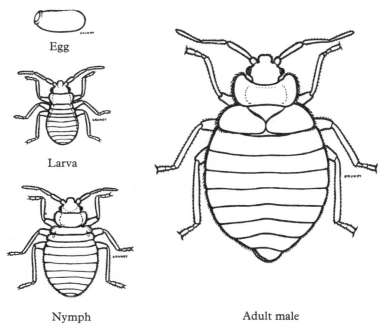

Egg

Larva

Nymph   Adult male

Figure 13   Bed-bug, *Cimex lectularius*

Life-history of Bed-bugs. The eggs of bed-bugs are quite visible (about one millimetre long), and are whitish and somewhat sausage-shaped (Fig. 13). They have a terminal cap or operculum by way of which the larva emerges, and there is a fine net pattern over the shell. The eggs are laid singly and intermittently over a period of several months, being glued behind wall-paper, or in crevices in furniture or walls. An average number of two hundred is laid by the female. The eggs hatch in about a week, depending on the temperature. The larvae and nymphs, which resemble the adults in miniature, grow rather slowly through their five moults, taking about seven weeks from egg to adult under favourable conditions. Some extra days are required before the adults begin to feed. The life cycle may be as short as a month or very much longer. The adults may live a year without food though the normal life-span is about six to eight months. As bed-bugs will readily feed on mice, bats, or birds, infested premises may remain so for years, even if uninhabited from time to time. Bed-bugs are less active in cold weather, but will usually feed if a host is available. There are normally about four generations a year.

Modes of Bed-bug infestation. Bed-bugs usually enter new premises in furniture or bedding which contain eggs, nymphs, or breeding adults. Migration from adjacent property is not uncommon. Poultry houses are sometimes infested with bed-bugs which will readily change their diet from the blood of fowls to that of man.

Control of Bed-bugs. Serious infestations with bed-bugs do not develop where the standards of cleanliness are high. Control is by the use of residual sprays on walls, floors and furniture. The persistent effect of these insecticides helps to prevent re-infestation.

Treatment with residual spray is effective for up to six months in temperate climates and two to three months in tropical and sub-tropical climates. Mattresses should be sprayed on both sides. Floors must not be scrubbed after treatment but may be washed five days after the spraying; they may be swept and dusted at any time.

CHAPTER 5

LICE OF MEDICAL IMPORTANCE

Order *ANOPLURA*

(Sub order SIPHUNCULATA)

Lice are always wingless and of small size, not usually exceeding one-eighth of an inch long. Under a hand lens the extremities of the legs will be seen to resemble the seizing-claws of crabs and lobsters. No other six-legged arthropods have legs of this type, which in lice are designed to grip hairs and clothing fibres.

Human Body and Head Lice *(Pediculus humanus)*

These lice may be found all over the world as parasites on the skin or in the hair or clothing of man. The variety found on the human head differs slightly from the larger more active variety found on the body or in the clothing, but the differences are not constant enough to separate head and body lice as distinct species.

Diseases caused and transmitted by Body and Head Lice.— 'PEDICULOSIS'. The bites of human lice are not usually painful, but they may cause a dermatitis which becomes aggravated by scratching. This may lead to secondary infections such as impetigo, infective dermatitis, and furunculosis. In persons who harbour body lice for years, the skin becomes roughened, thickened, and deeply bronzed by pigment. This condition is sometimes termed 'vagabond's disease'. Biting at night causes loss of sleep and may lead to neurasthenia.

Classical epidemic typhus. This is an acute infectious disease caused by certain rickettsiae and transmitted to human beings by body and head lice. Man is the reservoir of the disease. The faeces of a louse may become infective two days after a blood meal on a suitable typhus victim although the average incubation period is

53

five to nine days. Once infected, the louse remains so for the rest of its life, which is only about twelve days, as the multiplication of the rickettsiae intracellularly is fatal to the louse.

Relapsing fever. This is an acute infectious disease caused by certain spirochaetes. It is typically epidemic. The causative organisms make their way to the blood-space (haemocoele) in the body of a louse, after it has fed on the blood of a relapsing fever case. As the living spirochaetes do not remain in the intestine of the louse, and as they are not passed out in its saliva or faeces, man only becomes infected when he crushes the insect and releases the organisms on to his skin. A louse becomes infective some four to six days after the initial blood meal, although spirochaetes may appear in the louse-blood after some two hours or so. It remains infective for life, but does not pass the organisms to the next generation.

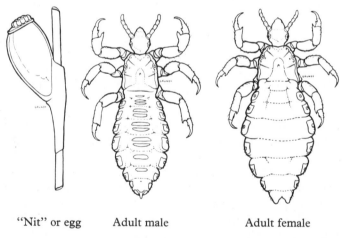

"Nit" or egg   Adult male   Adult female

Figure 14   Human louse, *Pediculus humanus corporis*

Recognition of Body and Head Lice. They are small (less than one-eighth of an inch long), grey with darker markings, flattened from above downwards and relatively elongate compared to crab lice (Fig. 14). They are without wings, and remarkable for the

54

toughness and elasticity of their integument, which is not strongly chitinous as in fleas. The head is distinct and longer than broad, with a pair of antennae composed of five segments. The head also bears a pair of simple pigmented eyes. The thorax is distinct and the segments are fused. The hair-gripping claws on the six legs are characteristic. The abdomen is distinct and somewhat heavy and long. There are characteristic indentations along the sides between the brown chitinous plates around the spiracles on the abdomen. The male louse has a rounded posterior contour and the female a forked posterior extremity. Head lice are rather smaller than body lice.

Habits of Head and Body Lice. All blood sucking lice are host specific, *e.g.*, human lice *(Pediculus humanus* and *Phthirus pubis)* are only found on man. Head lice are typically found on the human head, body lice on the body or clothing. Body lice are restless, and given to wandering about on the person. Though relatively slow movers they can cover about nine inches in one minute at room temperature. They are particularly fond of lurking in the seams of clothing. Lice feed frequently at all stages and do not usually live more than two or three days when separated from their human host. The maximum period they can survive away from man is about ten days. Lice will leave a dead body, or persons with an abnormally high temperature. This is important during epidemics. Male lice, if over numerous, will so worry the females that an individual infestation may die out. Lice readily sham death when disturbed, by lying still with their legs tucked beneath them. Head lice are less resistant to adverse conditions, such as partial starvation, than body lice.

Life-history of Body and Head Lice. The fertilized females lay individual eggs on hair or clothing fibres (sometimes several on one hair). These relatively large, easily visible white oval-shaped eggs (about the size of a pin's head), are attached to the hair or fibre by a large drop of gum-like cement. Under a hand lens the 'nits' as the eggs are called, may easily be differentiated from the eggs of other insects such as those of the bed-bug, by the presence of this solidified drop of cement at their base, when in conjunction with a cluster of pierced nodules partly covering a lid or operculum at the upper end of the egg. The developing larval louse breathes air through apertures in the nodules, and finally emerges from the 'nit' by forcing off this lid. It takes from five to ten days for the larva to hatch, and the young louse must feed on human blood within 24 hours if it is to survive. The eggs may remain dormant and viable for over a month if separated from the warmth of the host. There are two nymphal stages (*i.e.*, a larval and a first and second nymphal stage). Both larvae and nymphs resemble the adults in miniature. In a minimum of a fortnight, or an average of three weeks after the eggs are laid, adult forms will emerge and within 24 hours may feed, mate and even oviposit. The female louse lives about a month when on the host—the males do not live quite so long—and during this period she lays some 250 to 300 eggs, not all of which hatch. About ten eggs or less are laid in one day.

Crab Lice *(Phthirus pubis)*

On infested persons these lice are found invariably on the pubic hairs, but infestation may extend to the chest hairs, the armpit hairs and even to the eyebrows and eyelashes.

Diseases caused by Crab Lice. The term PEDICULOSIS is also applied to the skin condition caused by heavy infestations of crab lice. The habit of continuously feeding at one point on the body, together with the heavy contamination with louse faeces, and the complications caused by scratching, may cause a severe dermatitis. The host's skin may show a characteristic blue-grey discolouration due to continuously injected louse saliva. The crab louse has not been proved to transmit typhus or relapsing fever.

Recognition of Crab Lice. Under magnification (Fig. 15), the resemblance of these small squat lice to a sea-shore crab is very noticeable. The head is distinct, but the thorax and abdomen are fused so that the body appears heart-shaped and crab-like. Though the claws of the first pair of legs are weak, the claws and adjacent leg segments on the hinder two pairs are more powerful than any of the leg endings of the body louse. Crab lice anchor themselves to the pubic hairs at one spot and remain there for long periods with the mouth-parts inserted. They withdraw these, however, while

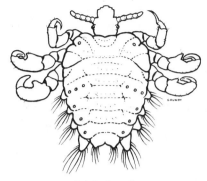

Adult female

Figure 15   Crab louse, *Phthirus pubis*

moulting. There are three spiracles in a dorsal transverse row on either side of the body about half way across its length. This arrangement is diagnostic. The appearance of these lice on the human body is that of specks or small areas of grey dirt attached to the hairs.

Life-history of Crab Lice. This is similar in essentials to that of the head and body lice. The females, however, only glue about 30 'nits' to supporting hairs. The eggs may be recognized by the lid or operculum being entirely covered by relatively elongate pierced nodules. Larval and nymphal crab lice closely resemble the adults in miniature. The life cycle is about two weeks.

Control of Lice. In conditions where a high standard of personal hygiene prevails, body and crab lice will not flourish; unfortunately, this is not entirely true of head lice. Control measures must always aim at the provision of good facilities for bathing, laundry and frequent change of clothing, particularly underclothing. The usual laundering methods destroy lice and their eggs, except that woollen materials may not be washed in sufficiently hot water.

In situations of natural or man-made disaster, conditions are such that the spread of louse infestation is greatly favoured, and control by general measures may be impossible. The use of modern insecticides, however, has made the control of louse infestation practicable.

Crab-lice are easily destroyed by dusting the affected area with anti-louse powder. Dusting should be repeated in a week to kill lice which may have hatched since the first treatment.

Anti-louse powder, when dusted into the clothing, is effective against body-lice. The great difficulty is to prevent re-infestation and therefore when numerous cases are occurring, or if bathing and laundry facilities are poor and conditions favour infestation, dusting of the clothing should be carried out weekly.

Head lice may be effectively controlled by the use of an anti-louse shampoo or preferably lotion.

Dog Lice (*Trichodectes canis*) (*Sub order* MALLOPHAGA). A quite different type of louse is found on the dog. This cream coloured, broad-headed, nibbling (not piercing and sucking) louse may devour such eggs of the dog tapeworm as adhered to the dog after being passed *per anum*. If a louse containing cysticerci from these eggs is ingested by a dog when licking itself or by a person handling the animal, infestation with dog tapeworm (*Dipylidium caninum*) is likely to occur in the dog or the human being respectively.

CHAPTER 6

FLEAS OF MEDICAL IMPORTANCE
Order *SIPHONAPTERA*

These fleas are wingless and of small size (about $1\frac{1}{2}$ to 4 mm.); they are always brown in colour and are the only arthropods of medical importance to be markedly flattened from side to side (the female jigger flea when swollen with eggs is the one apparent exception). For practical purposes, fleas are divided here into those which run freely on the host and those which attach themselves by their mouthparts for long periods.

'Free Running Fleas'

These active fleas are found all over the world as ectoparasites of birds and mammals. They have well developed legs for jumping upon a passing host, while their laterally flattened bodies are specially adapted for progress between hairs or feathers. Although not as host-specific as sucking lice, a species of flea may usually be associated with a restricted number of animal hosts.

Diseases caused and transmitted by 'running fleas'. The effect of the BITES of these fleas varies considerably in individual persons. In some there is immediate pain together with the formation of rose-red papules which may develop into pustules liable to secondary infection. In other persons no reaction will be discernible. Hungry fleas are known to have both predilections for and antipathies against certain individuals. The habit of restlessly running over the human skin and occasionally plunging in the proboscis for a blood meal may cause serious loss of sleep in an infested person.

Cestodiasis. Some fleas are the intermediate hosts and transmitters of the dog tapeworm *(Dipylidium caninum)* and the rat tapeworm *(Hymenolepis diminuta)*. The flea larvae ingest the worm

59

eggs passed *per anum* by a dog or cat, or rat, and retain the cysticer-coid stage of the worm in their body cavity until they are adult fleas. If one of these infected adults is accidently ingested, tapeworm development may result after a period of about 20 days.

Flea-borne typhus, endemic or murine typhus. This is an acute febrile disease of relatively mild form and low mortality. It is caused by certain rickettsiae which are sometimes conveyed to man by fleas which have fed on infected rats. The ingested organisms enter the epithelial cells of the flea's mid-gut and multiply enormously. About ten or twelve days later the rickettsiae return to the lumen of the gut and may remain alive for seven weeks or more. The organisms do not harm the flea. Inoculation into man does not seem to occur through the saliva of the flea but rather by the rubbing into the bite-puncture of infected flea faeces or crushed flea bodies. Flea-borne typhus is most common in summer and autumn amongst male workers in docks or grain stores where rats are plentiful. Louse-borne typhus invariably occurs in the winter months among persons of both sexes who are crowded together in unhygienic conditions.

Bubonic plague. This is an infectious and at times a contagious disease with an unusually high mortality. It is caused by certain bacilli which are transmitted to man by fleas which have previously

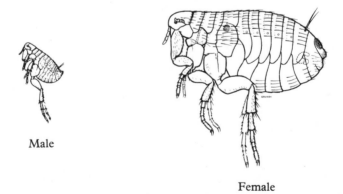

Male

Female

Figure 16   Plague flea, *Xenopsylla cheopis*

fed on infected rats or human cases of the disease. The chief mode of transmission is believed to take place *via* the mouthparts of the flea, particularly in those fleas which have become 'blocked'. This occurs when the entrance valve (proventriculus) to the flea's mid-gut has become impassable through the enormous multiplication of plague bacilli. Such fleas often pass restlessly from host to host in an endeavour to obtain blood to assuage their thirst. By regurgitation (or occasionally by infective faecal deposits), the bacilli find entrance from the 'blocked' flea to the bite puncture or other abrasion. The main reservoir of plague is in the rat population of endemic areas. Endemic rodent plague in wild rodents, other than domestic-type rats, in country and forest areas is sometimes termed sylvatic plague. The maintenance and spread of plague is dependent on chains of complicated circumstances, which carry the disease to epidemic proportions at periodic intervals. *Xenopsylla cheopis*, the common tropical rat flea, is the most easily 'blocked' flea in endemic areas, and so the most notorious vector of plague.

Recognition of 'running fleas'. Adult fleas (Fig. 16) are easily recognized by means of a hand lens, although when seen with the naked eye they are occasionally mistaken for other creatures on account of their small size. They are always brown in colour and shiny, being completely clad in a series of hardened smooth chitinous plates. There is no visible neck separating the head from the thorax (as in flies), and the antennae are segmented club-like organs lodged in pits behind the eye position. The eyes, when present, are always simple (not compound). These points are characteristic of all fleas. In 'running fleas' the hind legs are always powerfully adapted for running and jumping, being especially developed with very large coxae. The coxa is that leg-segment nearest to and adjoining the body, and in most insects is a comparatively small segment. The three thoracic segments are necessarily well developed for articulation to the three pairs of powerful legs. Female fleas which are not too dark in colour show under magnification a curious comma-shaped organ, within the posterior portion of the abdomen, known as the spermatheca. The shape of this is useful in the differentiation of species.

Xenopsylla cheopis. This notorious plague vector is a rather small,

61

light-chestnut coloured flea. It increases a little in size and darkness of colour after feeding, and the female (as with most fleas) is larger than the male. It is a common flea of rats in the tropics.

Identification of the genus *Xenopsylla* under the microscope is carried out by examination for a pair of eyes, together with the presence of a pair of brown structural 'vertical bars' which are situated within the sternite of the mid-thoracic segment, immediately above the junction of the second pair of legs. Although other genera of fleas may also have 'eyes' and 'bars', they will display differentiating additions such as 'combs' (broad blunt flat bristles which are socketed to the edge of a segment in a row), or 'teeth' (sharp unsocketed projections of the chitin) on the cheek (gena). The absence of eyes (as in *Leptopsylla segnis* the mouse flea) or absence of 'bars' (as in *Pulex irritans* the human-flea) at once designates the flea as '*not a Xenopsylla*'.

Habits of 'running fleas'. While living as blood-sucking ectoparasites on their several hosts, fleas often leave the animal's body and wander into the material forming its bedding. Fleas may thus be found deeply buried in the wool of blankets. They may be detected by holding up the article to a strong light. In the event of their usual host dying, fleas will transfer themselves to the nearest warm-blooded animal. The dissemination of uninfected fleas is chiefly by means of the eggs being widely scattered, with consequent attachment of the future adults to hosts which are likely to be other than those parasitized by their parents. 'Running fleas' feed frequently, usually attaching their mouthparts to the host for short periods several times a day. Once they are used to feeding, however, they do not easily withstand starvation, or a high dry temperature. Resting unemerged adults, on the other hand, can remain in their cocoons for months, awaiting the vibration of a passing host on which or whom they will endeavour to jump. While imbibing a blood meal, this type of flea habitually evacuates considerable quantities of almost pure blood. This dries on the hosts' skin or hair, and when rubbed off drops to the ground and forms a main food supply for the larval fleas. *Xenopsylla cheopis*, being a small flea, can only hop five inches, but *Pulex irritans* the human flea, is capable of a horizontal leap of 13 inches and a high jump of almost 8 inches.

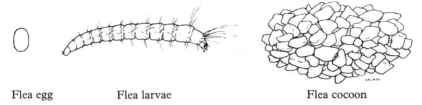

Flea egg   Flea larvae   Flea cocoon

Figure 17

Life-history of Fleas. The life history of all fleas is essentially similar (Fig. 17). The times of development and situations of the habitat, however, vary considerably. The gravid female flea deposits several ovoid pearly white eggs at one time, while still on the body of her host. As these eggs lack an adhesive cement, they usually drop to the ground or floor. Their appearance on a dark background is not unlike grains of white sugar. After two or more days, depending on temperature and humidity, the eggs hatch into small elongate legless maggots, which wriggle actively by means of concentric rows of radiating bristles, and movements of the head. They bury themselves in any dark crevice where organic debris may be found. A diagnostic feature is the pair of blunt ventrally pointing anal struts at the posterior end of the larva. There are three larval stages which may take a minimum period of seven days *(Xenopsylla cheopis)*. The third stage larva finally attains the resting or pupal period after enclosing itself in a silk cocoon to which adheres a sufficient quantity of the local dust and debris to act as camouflage. The pupa may develop into the adult flea in about a week. This adult will not necessarily emerge from its protective covering, however, as it usually waits for a passing host to vibrate the cocoon, before pushing aside the silk strands and leaping vigorously towards the cause of the disturbance. A house left uninhabited for months may in this manner greet new occupants with hundreds of hungry fleas.

A female *Xenopsylla cheopis* under suitable conditions may lay between 300 and 400 eggs. Her life may be a little over three months if she is regularly fed and a little over a month if fed once only. The average life cycle for this flea is about three weeks. Other species of fleas in a colder environment may take a year for their life cycle. A minimum period of two weeks, however, may be given.

63

'Sticktight Fleas'

Tunga penetrans, the 'jigger flea' is distributed in a drier habitat than is usually acceptable to the 'running fleas', since it is normally found in dry and sandy soil in and around native huts, stables and pigsties in Central and Brazilian Neotropica, and Tropical and Madagascan Ethiopia. It has been found in India and the U.S.A.

Disease caused by Jigger Fleas. TUNGIASIS.—The fertilized females burrow into the skin of the host. In human beings beneath the toenails is a favourite position. Here the abdomen of the flea distends with blood and developing eggs until it is the size of a small pea. Intense itching and inflammation, and secondary infection and ulceration, not infrequently result. Amputation of the toes or the lower limb may be necessary. Numerous deaths have occurred owing to complications of gas gangrene and tetanus resulting from jigger-flea infestation.

Recognition of Jigger Fleas. These are very small fleas which appear as little dark specks less than a twenty-fourth of an inch long (Fig. 18). With the aid of a good hand lens the characteristically pointed forehead (frons) can be made out. Since jigger fleas tend to attach themselves by their mouthparts and not wander about the host, their legs are not so strongly developed as are those of the 'running fleas'. Neither is the size and development of the thoracic

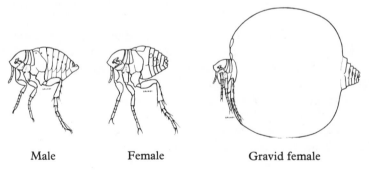

Male   Female   Gravid female

Figure 18   Jigger flea, *Tunga penetrans*

64

segments so marked, for these are so narrow and pressed together as to be described as 'telescoped'. This is most noticeable in the pregnant female when her head is tightly compressed against the tissue of the host by the swelling of her own body within the restricted space of her 'burrow'. The male jigger always remains a tiny, free-living, somewhat weakly built flea. The female, however, after fertilization and burrowing, swells to such a large size by distension of the membrane between her second and third abdominal segments as to be quite unmistakable. The clinical appearance of the lesion is that of a small black spot (the abdominal segments of the flea), in the centre of a tense rather pale area of swelling.

Habits of Jigger Fleas. The adults feed on serum and perhaps the blood of their bird and mammalian hosts. The males are free living, as are the females before fertilization, and lead a somewhat similar life to the 'running fleas'. After becoming fertilized, however, the female jigger seeks a host and, attaching herself by means of her powerfully serrated mandibles, burrows beneath the skin until only her posterior abdominal segments and spiracles are visible. The principle hosts seem to be pigs and man, but cats, dogs, rats and domestic fowls are also infested. Jigger fleas display a strong preference for a sandy habitat.

Life-history of Jigger Fleas. The ovaries of the burrowing female develop rapidly and their increase in size soon fixes her firmly in position within the lesion. When fully gravid the eggs are forcibly expelled and being somewhat sticky tend to adhere to whatever material they touch. Usually this is the dust of the habitat. A hundred or more eggs are laid in all. The larvae are similar to those of other fleas and after feeding on such organic debris as can be found, and moulting three times, they change into pupae inside a silk cocoon camouflaged with sand or dust. After a few days quiescence the adult jigger fleas are ready to emerge.

Control of Fleas. Infestation is unlikely to occur where standards of domestic hygiene are good and where regular airing and washing of bedding are carried out. Cleaning and sweeping, particularly with the vacuum cleaner, are inimical to fleas.

In the United Kingdom, except in insanitary surroundings, infestations are likely to be due to domestic animals such as dogs

65

and cats. Kennels and baskets must be kept clean. Infestation may be cured by the use of residual sprays which prevent re-infestation.

Overseas, where there is a risk of plague, rodent control is intimately connected with measures against fleas and is particularly important.

Those who have to deal with heavy infestations should wear protective clothing, such as gloves and gum-boots or boots and anklets. Repellents are effective. They may be applied to the skin or smeared over the clothing in such areas as the socks, trouser-ends and cuffs. Such application to the clothing should remain effective for 10 to 14 days.

Anti-louse powder dusted into the clothes as for lice is effective in destroying fleas which have gained access to the body. The occasional flea may easily be caught in soapy hands.

Infestations can readily be cleared up by the use of residual insecticides, applied to floors, furniture and the lower three feet of walls. These insecticides are also effective when applied in powder form. The corners of rooms, rat-holes and runs should have particular attention. It is important, when powders are used, that floors should not be swept or scrubbed for several days after treatment, which may have to be repeated as eggs hatch out.

The control of flea-borne plague has been simplified by the use of modern insecticides. The application of insecticide powders to the immediate neighbourhood of the case, including all rat runs and holes, destroys infected fleas and gives rapid control. The destruction of rats must, however, still be relied upon for the long-term control of plague.

Protection against the jigger (sand-flea, chigger or chigoe) is given by the wearing of boots and shoes. Walking barefoot and sleeping unprotected on the ground should be avoided, and also camping in or close to, infested areas. The use of anti-louse powder on the floors of huts and tents will give protection, as will the use of repellents (see above). Daily inspection of the feet for signs of infestation is important.

CHAPTER 7

FLIES OF MEDICAL IMPORTANCE

Order *DIPTERA*—(*dis*, twice; *pteron*, wing)

These insects typically have two wings, *i.e.*, they have one pair of flying-wings. A vestigial pair of hind wings has been modified into balancing organs, or 'halteres'. They are all included in the general term 'flies'. There is a distinct neck between the head and thorax, and a waist between the thorax and abdomen. A pair of antennae and three pairs of walking legs are always 'present. The mouth-parts are adapted for sucking fluids and *not* for gnawing. The immature stages are 'grubs' or 'maggots' and are quite unlike their parents. There is always a non-feeding quiescent pupal stage, wherein the tissues of the larva break down and rebuild into the winged fly (= complete metamorphosis). Flies are highly specialized and successful in the battle of life, and the number of different species is legion. In order to find out if a two-winged fly belongs to a genus or species of medical importance, it is a useful plan to examine the antennae through a hand lens, and so become used to narrowing down specimens to one of three groups or suborders:

(1) Antennae long and thin, and of not less than six essentially similar segments together with two dissimilar basal segments (*i.e.* a minimum of eight segments) NEMATOCERA, including sandflies, mosquitoes, biting-midges, and blackflies.

(2) Antennae generally shorter and stouter than in Nematocera, and typically of three dissimilar segments, the third sometimes being ringed or annulated to appear as several segments BRACHYCERA, including the gadflies.

(3) Antennae short and stout and of three dissimilar segments, with a dorsal bristle or arista on the enlarged third segment ATHERICERA, including the houseflies, blow-flies, and bot-flies.

Suborder NEMATOCERA (*nema*, thread; *keras*, horn or antenna)
SANDFLIES (*Phlebotomus*)★. These tiny flies are found in warm and

★*Phlebotomus* in Old World; *Lutzomyia* and *Sergentomyia* in New World.

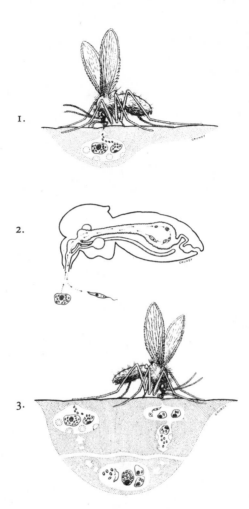

1. Sandfly ingesting Leisman-Donovan bodies from human host.

2. Development cycle from L.D. bodies to infective leishmaniae within gut of fly. This takes about seven days. Development is broken if blood meal occurs about fifth day, but is accelerated if sweetened fluid is taken instead.

3. Injection of infective leishmaniae into new host during subsequent blood meal. Organisms survive best in tissue macrophages.

Figure 19   Leishmaniasis cycle in man and the *Phlebotomus* sandfly

tropical regions in coastal and mountainous areas. They are absent from the colder parts of temperate zones.

Diseases caused and transmitted by Sandflies. The BITES of sandflies are immediately painful and produce an irritating pruritis or itching which is similar in appearance to chickenpox. In some cases this condition may last for weeks and even develop into a marked systemic toxaemia and fever. Immunity is often acquired after a moderately long association with the bites of sandflies.

Sandfly fever. This is an acute fever of three days' hyperpyrexia, and is caused by a virus which is capable of being transmitted by certain species of sandflies, some seven to ten days after they have fed on a human sufferer during the first 24 hours of the illness. A single sandfly bite may transmit the disease. The period of incubation in man is approximately four to eight days. The eventual prognosis is good. Sandfly fever is endemic in the Mediterranean Palaearctica and is also prevalent in the Oriental region. A known vector is *Phlebotomus papatasii.*

Oroya fever or Carrion's disease. This is an acute and sometimes fatal disease caused by bacillary organisms and transmitted by certain species of sandflies in certain high valleys (2,500 ft. to 8,000 ft.) in the Andean Neotropica. A known vector is *Phlebotomus (= Lutzomyia) verrucarum.*

Leishmaniasis. VISCERAL LEISHMANIASIS or kala azar is a long-duration fever caused by *Leishmania donovani* and transmitted to man by the bites of certain species of sandflies (Fig. 19). Man is the principal reservoir, and the disease is endemic in the Mediterranean and southern Mongolian Palaearctica, the eastern part of North Tropical Ethiopia and the Oriental region. Sandflies which have the pharynx 'blocked' with leptomonad forms, following a meal of plant juices some days after the infected blood meal, are the dangerous vectors.

Phlebotomus argentipes (India), *Phlebotomus chinensis* (China), *Phlebotomus perniciosus* (Mediterranean), are considered important vectors.

Cutaneous leishmaniasis or Oriental sore is a specific granuloma of the skin which usually breaks down to form an indolent ulcer.

69

It is caused by *Leishmania tropica* and is transmitted by certain species of sandflies. It is endemic in extensive areas in the Mediterranean and Mongolian Palaearctica, and in the Ethiopian and Oriental regions. The sandflies become infective to man several days after a blood meal from the area around the sore. The chief reservoir is man, but other mammals have been found infected in nature.

Phlebotomus papatasii and *Phlebotomus sergenti* (India) and *Phlebotomus macedonicum* (Italy) are considered vectors.

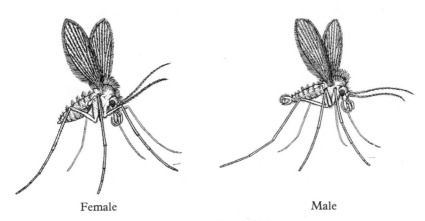

Female   Male

Figure 20   Sandfly, *Phlebotomus*

Recognition of Sandflies. The adult flies are extremely small, less than 3.5 mm., so that they easily pass through ordinary mosquito netting (Fig. 20). They are covered with a dense hairy coat of a sandy yellow or brown colour. The eyes are conspicuously dark. The antennae are very long and of 16 segments. The beak is dagger-like and points ventrally. The palps are recurved below the head, while those of the mosquito project forward.

The thorax is markedly humped and the legs are rather long. Under a microscope the wing venation, when denuded of hairs in a little fluid, shows the *second* longitudinal vein to be forked twice, once about the middle of its length and again towards the wing-tip. The male sandfly has prominent terminal organs for clasping the female.

70

First-stage larva   After first-stage larva

Figure 21   Sandfly *(Phlebotomus)* larva

Habits of Sandflies. These tiny flies usually rest in dark damp places, and when disturbed make characteristic little hops, which are *not* to be confused with the jumping of fleas. Sandflies are able to fly reasonably well, though the usual radius is about 50 yards from the breeding place. It is believed that only the female sandflies are blood sucking, since blood is necessary for development of the eggs. Usual feeding time is at dawn or dusk or during the dark hours. The approach is silent, unlike that of most mosquitoes, and the attack is usually upon exposed parts such as the neck and chest, arms and lower legs.

The Life-history of Sandflies. The females of some species feed but once and die after ovipositing. Other species feed several times and lay batches of eggs between the blood-meals. The eggs are minute and somewhat elongated. They are laid in cracks and crevices in dark moist sites, frequently at the base of buildings where there is shade. The eggs hatch in nine to twelve days and there are four larval stages and moults, followed by a pupal stage. The pupa retains the characteristic fourth larval skin on its posterior extremity. The first stage larva has *two* very long bristles projecting from its caudal extremity (Fig. 21). The second, third, and fourth stage larvae have *four* long bristles projecting from the caudal segment. These bristles, together with the many short club-shaped body hairs are diagnostic of the genus. The larvae eat decaying organic matter. The life-cycle under favourable conditions takes place in about one month, but may be considerably lengthened under adverse circumstances.

Control of Sandflies. Control has been made easier by the use of modern insecticides. Other measures, however, still have a place. Such are the use of insect repellent and the wearing of protective clothing (slacks and long sleeves) to diminish the area of skin available to the sandfly for biting. Free air movement produced by

71

fans or natural ventilation is useful for discouraging the sandfly which is a weak flier. A mosquito net which has been impregnated with repellent protects at night; sandfly nets, which are uncomfortable because of their very small mesh, are no longer used. Sleeping quarters are better situated on the first floor or higher because sandflies are much less numerous above the ground floor.

Elimination of breeding-places within a radius of 250 yards will greatly diminish the number of sandflies coming into a building, because of their limited range of flight. Places which afford the necessary darkness, moisture and organic matter are dealt with by such measures as the removal of rubble, levelling the ground by rolling, facing and pointing walls to fill in crevices, filling in cracks in banks of streams and drains with concrete or cement, and, in special cases, rendering the ground surface impermeable with cement, asphalt, tar, or similar material.

Such measures have become unimportant compared with the use of residual spray. The treatment of interiors with residual spray gives immediate and almost complete protection from sandflies indoors. The spraying of interiors alone will eventually reduce the sandfly population of a particular area almost to vanishing point; this effect is best shown in built-up areas.

Spraying of resting and breeding-places with residual spray for an area of 250 yards around buildings is a useful supplement to indoor spraying. Although the eggs are not affected, emerging sandflies will be killed. The use of aerosols, such as anti-mosquito/anti-fly spray, for their quick knock-down effect, may be useful for clearing a room of sandflies at night.

Mosquitoes (Subfamily Culicinae)

These well-known flies may be differentiated from their near relatives by the presence of a long forwardly projecting proboscis or beak, together with a pair of wings of a characteristic pattern. As there are hundreds of species it is most useful to group any individual specimen into one of three tribes.

(1) Those with a downwardly hooked proboscis MEGARHININI large irridescent jungle mosquitoes, of no medical importance.

(2) Those with the abdomen typically banded with light and dark coloured scales (Fig. 22) CULICINI containing vectors of Yellow fever, Filariasis and Dengue fever.

(3) Those with the abdomen *not* banded with scales ANOPHELINI containing vectors of Malaria and Filariasis.

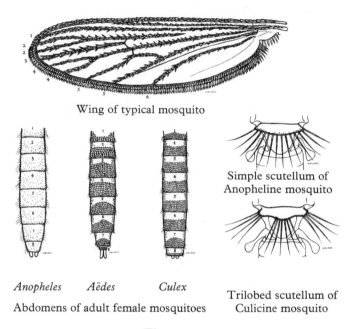

Wing of typical mosquito

Simple scutellum of Anopheline mosquito

Anopheles   *Aëdes*   *Culex*

Abdomens of adult female mosquitoes

Trilobed scutellum of Culicine mosquito

Figure 22

Species of either Anophelini or Culicini may, in different localities, display a preference for animal blood (*zoophilism*) or for human blood (*anthropophilism*). Of the anthropophilic mosquitoes there are 'domestic' species which breed and live near or in dwellings, and 'wild' species which breed and chiefly attack man in the open. The above habits are of medical importance.

Diseases caused and transmitted by Mosquitoes. The BITES are not usually painful but frequently produce a local reaction lasting for a few hours up to several days. This condition is heightened in some persons by allergic sensitivity.

73

Malaria. The invasion of the blood of man by malarial plasmodia is due entirely to the ability of the females of certain species of Anophelini to incubate the pathogenic organisms through part of their life-cycle (sexual stage) and transmit the developed forms to a new human victim by bite. This disease, of which the endemic reservoir is man, occurs over half the world's land surface, wherever conditions are suitable for the vectors to breed and become infective. The following mosquitoes are dangerous vectors of malaria:

Anopheles quadrimaculatus (N. America), *Anopheles albimanus* and *Anopheles darlingi* (Tropical America); *Anopheles maculipennis labranchiae*, *Anopheles maculipennis atroparvus*, *Anopheles saccharovi* and *Anopheles superpictus* (Europe); *Anopheles claviger*, and *Anopheles sergenti* (North Africa and Middle East); *Anopheles funestus*, *Anopheles gambiae* and *Anopheles melas* (Tropical Africa); *Anopheles stephensi* (Middle East); *Anopheles culicifacies*, *Anopheles fluviatilis*, *Anopheles minimus*, *Anopheles sundiacus*, *Anopheles varuna* (India and Ceylon); *Anopheles maculatus* (Far East); *Anopheles punctulatus* (Australasia).

Yellow fever. This is an infective fever caused by a virus which is transmitted to man chiefly by a 'domestic' culicine mosquito *Aëdes aegypti* (Fig. 23). The endemic reservoir of the disease is in jungle monkeys and other animals, and is perpetuated by various 'wild' chiefly day biting culicine species. If a person visiting these jungles is bitten by an infected 'wild' mosquito, he may return to a populated area where the domestic *Aëdes aegypti* is breeding, and start an epidemic of the disease. Species of the genera *Aëdes* and *Eretmopodites* in Africa and of the genera *Aëdes* and *Haemagogus* in America are the main vectors of jungle yellow fever.

Filariasis (mosquito-borne). This helminthic disease is caused by two species of filarial worms (*Wuchereria bancrofti* and *Brugia malayi*) which may infest and block the lymphatic channels of man and so cause enormous swelling of parts of his body. It is transmitted by certain species of Culicini and Anophelini. *Culex fatigans* is perhaps the most notorious vector. The mosquito first imbibes the microfilariae from the blood of an actively infected person. Under optimum conditions the mature larval worms, which have meanwhile undergone development within the body of the insect, find their way about the tenth day into the blood space (haemocoele) of the proboscis or beak-sheath of the mosquito. By rupturing the wall of this, they escape into the bite-puncture or bore through the skin of the human being on whom the mosquito is about to feed. The endemic reservoir of the disease is man in many warm regions of the world. A large number of *Anopheles, Aëdes, Culex* and *Mansonia* species are concerned in transmission.

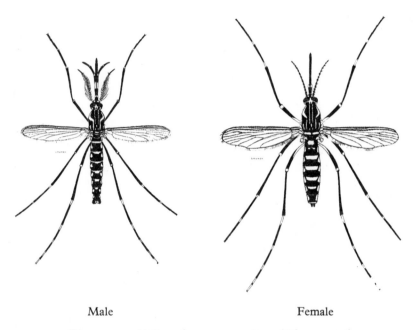

Male   Female

Figure 23   Yellow fever mosquito, *Aëdes aegypti*

Dengue fever. This is an acute infectious fever which invariably produces complete temporary incapacity but which is rarely fatal. It is caused by a virus which occurs endemically in native popula-tions of most warm countries, but it may develop wherever the 'domestic' culicine *Aëdes aegypti* or other *Aëdes* vectors breed. The mosquito becomes infected by feeding on a person during the first three days of the illness, and after an approximate incubation period of eight to eleven days is able to transmit the virus to man by

75

bite. Once infected the insect remains so for life. *Aëdes albopictus* and *Aëdes scutellaris* are known vectors.

Recognition of Mosquitoes. Adult mosquitoes are slender two-winged flies, varying in size from about an eighth of an inch to about half an inch in length. They are of a brownish colour, though typically decorated with patches of cream, white or dark blue or brown scales. The three pairs of walking legs are long and thin, this being particularly noticeable in Anophelini. The wing design is diagnostic. Under a hand lens the wing is seen to be long and narrow, and rounded at the apex. The third longitudinal vein is short and simple, at the tip of the wing, and between two forked veins. All the veins have scales and the hind margin of the wing has a distinct fringe of scales. Since it is only the female mosquitoes which pierce the skin and suck blood, it is necessary to be able to recognize the sexes, if examination of specimens for pathogenic organisms is undertaken. Male mosquitoes have bushy (plumose) antennae. Female mosquitoes have sparsely-haired (pilose) antennae. In case of doubt as to the 'banding' or 'not banding' of the abdomen with scales, as when a specimen is shrivelled or damaged, there are also useful secondary characters to be seen with a good handlens or a low power microscope.

ANOPHELINI	CULICINI
(containing one genus *Anopheles*) (Fig. 24).	(containing many genera, *e.g.*, *Aëdes, Culex, Mansonia*)
The palpi of the males are as *long* as the proboscis and terminally '*clubbed*'.	The palpi of the males are as *long* as the proboscis and are *not* '*clubbed*', although they may be much ornamented and feathery.
The palpi of the females are as *long* as the proboscis and *simple*. The scutellum has a *simple* posterior outline with an evenly spaced row of backwardly pointing bristles in an undamaged specimen.	The palpi of the females are *very short* and simple. The scutellum has a *trilobed* posterior outline, with a backwardly pointing tuft of bristles on each lobe in an undamaged specimen.

76

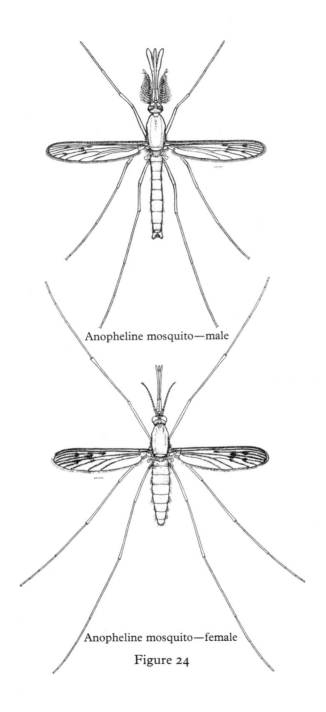

Anopheline mosquito—male

Anopheline mosquito—female

Figure 24

The scutellum is easily distinguished.	The scutellum is difficult to distinguish, because it is partially overlapped by thoracic hairs and scales.
Wings of many species 'spotted' with light and dark patches of scales.	Wings rarely spotted with light and dark clumps of scales.

There are a few exceptions to the above differentiations.

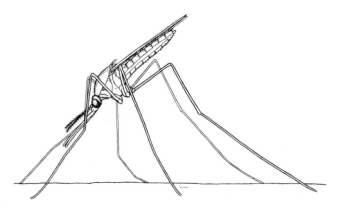

Culicine mosquito—resting position

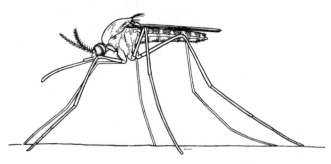

Anopheline mosquito—resting position

Figure 25

Habits of Mosquitoes. Both male and female mosquitoes will imbibe sweet juices from flowers or ripe fruit. Only the females pierce the skin and suck the blood of vertebrate animals. This is to complete the development of their eggs. The chief times for feeding are at dawn and dusk and to a lesser degree during the dark hours. Some species, however, will attack viciously by day. Most mosquitoes make a whining nose while flying. Some, such as *Aëdes aegypti*, attack more silently. When resting on a wall or other surface, mosquitoes have a habit of raising their long *hind* pair of legs into the air above their backs. These raised legs apparently act as feelers. The non-biting Chironomidae which are near relatives of the mosquitoes and often mistaken for them invariably raise the *front* pair of legs into the air for a similar purpose. Anophelini when resting usually tilt the abdomen away from the wall surface at an angle, *i.e.*, they appear partially to 'stand on their heads'. Culicini when resting, generally keep the abdomen parallel to the wall surface (Fig. 25). This is due to culicine mosquitoes having a much more humped thorax than anopheline mosquitoes. There are very few exceptions to this.

The female mosquito rarely takes more than a minute or so to engorge with blood and to transmit disease if she is infective. Mosquitoes are able to fly considerable distances, though the usual range of flight is not more than a mile from the breeding place. The longest flights occur during the mass migrations of some species, five, ten, and twenty mile flights not being uncommon.

The day is usually spent resting in a dark cool corner of a dwelling or in the shade of bushes, holes or caves.

Life-history of Mosquitoes. All mosquitoes have an egg stage, oviposition taking place on or near water; an aquatic larval stage, including four moults to enable growth to take place, and a non-feeding pupal stage which finally produces the adult winged form. All stages breathe atmospheric air. Due to slightly different feeding habits, Anophelini and Culicini, in their immature stages, are recognizably different in shape, though the pupae are somewhat similar.

Anopheles eggs

Anopheles eggs in 'rosette' formation

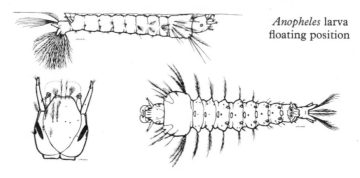

Anopheles larva floating position

Head of *Anopheles* larva   Anopheline larva (dorsal view)

Figure 26   Anopheline mosquito, Tribe Anophelini

ANOPHELINI (Fig. 26)	CULICINI (Fig. 27)
The eggs have lateral air-floats (An exception is *Anopheles saccharovi*).	The eggs never have lateral air-floats. They may be laid separately as in *Aëdes*, or glued in rafts as in *Culex*.
The larvae float and feed with the body *parallel* to the water surface. The breathing spiracles near the tail-end are therefore *flush* with the segments.	The larvae either *hang* from the water surface, or feed on the bottom. The breathing spiracles near the tail-end are therefore placed on a *projecting siphon* tube.

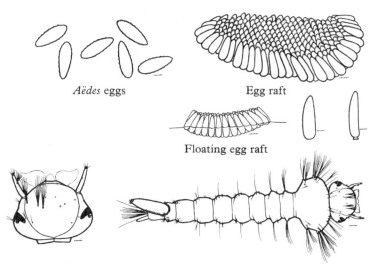

Aëdes eggs   Egg raft

Floating egg raft

Head of larva   Culicine larva (dorsal view)   Separate eggs from rafts

Figure 27   Culicine mosquito, Tribe Culicini

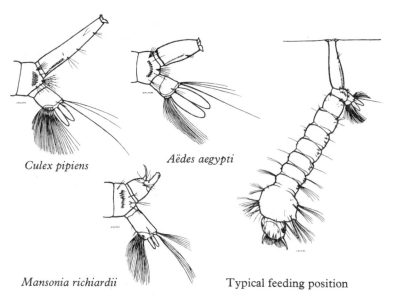

Culex pipiens

Aëdes aegypti

Mansonia richiardii   Typical feeding position

Figure 28   Breathing siphons of Culicine larvae

Mosquito larvae (Fig. 28), though very active swimmers or 'wrigglers', are heavier than water, and sham death on the bottom when disturbed. They feed either by sweeping the water with mouth brushes for organic particles, or by nibbling with their mandibles at dead leaves and the like.

The pupae (Fig. 29) of Anophelini and Culicini are very similar in general appearance. They are 'comma-shaped', with a flexible abdomen for swimming, and a pair of cone-shaped trumpets on the dorsal surface for breathing air. They are lighter than water and rest at the surface unless disturbed, when they may dive actively.

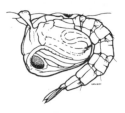

Resting position   Pupal paddles

Figure 29   Mosquito pupa

An exceptional method of breathing is that of *Mansonia* (which includes some vectors of filariasis). The larva pierces the stem of a rush or tall water-plant with its caudal breathing apparatus and obtains oxygen from the tissues of the plant. The pupa similarly attaches itself to plants by the trumpets.

Each species of mosquito seems to have a marked preference for a certain type of breeding place and is often unable to multiply if these sites are destroyed. The period of the life-cycle varies greatly in the different species, according to the temperature and the larval food supply. *Aëdes aegypti*, for instance, may have a minimum life-cycle of nine days or the period may be extended to many months by adverse conditions. Their larvae have a minimum growing period of six days and the pupae of one day. The average life-cycle for *Aëdes aegypti*, with good conditions of temperature and food supply is about eleven days to three weeks. A longer variation in the life-cycle of mosquitoes is usually due to a lowered temperature and a poor larval food supply or else to a specialized life-history. Some species pass the winter months in the egg stage, others in the larval, while others are known to hibernate completely or partially in the adult stage.

The control of Mosquitoes. The development of insecticides has brought about radical changes in tactics for the control of mosquitoes, both in the means employed and the emphasis on the different available methods. In malarious areas and other places where mosquitoes are a health hazard the approach to the control of mosquitoes is briefly summarized as follows:

(a) *Accommodation siting :* This is of great importance if a choice is possible.

(b) *Personal protection* is of fundamental importance because it can always be practised, the appropriate measures depending on the situation. The measures employed are: protective clothing, repellents, mosquito nets, and, where possible, screened quarters.

In addition, in malarious areas of the world anti-malarial drugs as recommended should be taken regularly to provide protection from the disease.

(c) *Residual spraying :* Residual spraying of quarters with residual insecticides ensures that a mosquito which alights on the sprayed surface is killed after a short time.

(d) *Insecticide sprays and aerosols :* These agents produce a rapid 'knock down' of adult mosquitoes. They are used in buildings to supplement residual spraying.

(e) *Larval control :* Measures to control mosquito larvae may be either temporary or permanent according to the length of time the area is to be occupied. A stay of only a few months calls for temporary measures; permanent measures may require several years to complete. If a site is not to be occupied for longer than a week or so, anti-larval measures will not have time to yield useful dividends.

Culicoides Midges

These are most troublesome outdoor pests of both temperate and tropical regions.

Diseases caused and transmitted by *Culicoides.* THE BITE.

83

Since *Culicoides* often attack in swarms and also creep between the head-hairs, they can be a considerable nuisance. The bite itself is felt as a sharp prick and is often followed by irritating lumps, which may disappear in a few hours, or last for days accompanied by some systemic toxaemia.

Filariasis. *Dipetalonema perstans* is the cause of an helminthic infestation of the human body cavities such as the peritoneal cavity and the pericardial sac. There are apparently no marked clinical manifestations. The endemic reservoir is chiefly man, but chimpanzees, other apes, and some monkeys are involved. The microfilariae are obtained from the peripheral blood of the sufferers by three species of *Culicoides*, which may later transmit the disease to man by bite. Endemic regions are the Mediterranean Palaearctica, Tropical Ethiopia, northern Neotropica and some parts of Polynesian Oceania. *Culicoides austeni* and *Culicoides grahami* (in Africa) are incriminated as vectors.

Mansonella ozzardi is also the cause of an helminthic infestation of the human body cavities. It produces no marked clinical reactions.

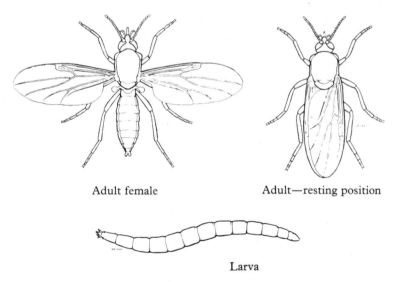

Adult female   Adult—resting position

Larva

Figure 30 Biting midge, *Culicoides*

The endemic region is restricted to northern Neotropica, and the reservoir is believed to be man. The worms are transmitted by *Culicoides furens* via the bite.

Recognition of *Culicoides.* The adults (Fig. 30) are minute dark coloured flies, rarely more than 2 mm. in length. The antennae are long (of 15 segments) and are plumose in the males and pilose in the females. Only the females bite and suck blood. Under a microscope the wings show two characteristic cells associated with the first long vein near the anterior margin. To the naked eye the wings show a characteristic spotting due to wing pigmentation.

Habits of *Culicoides.* The females typically bite at dusk, often in dense swarms and usually in the vicinity of water, marsh, or rotting vegetation.

Life-history of *Culicoides.* The eggs are minute and laid in a row or in batches on decaying vegetation in water or on land. Some species even breed in salty water. The larvae are tiny smooth whitish worms, which swim with a serpentine motion when disturbed. The pupa is a floating chrysalis-like object (not unlike a mosquito pupa) with a pair of breathing trumpets. The bionomics of these insects are but imperfectly known.

Control of *Culicoides.* Control is hindered by lack of knowledge of the habits and life history of the important species.

Simulium (Black-Flies)

These small biting flies are found in both temperate and tropical zones, in the vicinity of running water. They are particularly numerous in northern temperate and subarctic regions.

Diseases caused or transmitted by Black-flies. The BITES of *Simulium* may or may not cause a sharp pricking sensation, followed by pronounced itching, swelling, and ulceration. The eyes of the person attacked may become closed and the features temporarily distorted in appearance. This condition is known as 'bung-eye'. Fever, intestinal disturbance and incapacity may occur for weeks.

Filariasis. *Onchocerca volvulus* causes an helminthic infestation of the subcutaneous connective tissue of man and manifests itself in the form of tumours, with blindness as an occasional sequel. The disease is transmitted by some species of *Simulium*, which trap in their mouthparts microfilariae from the skin of an infected person, and after these have undergone development in the fly's thoracic muscles during some six days or so, pass the mature larval forms to a fresh human victim, while attempting to bite him. The larval worms rupture the beak or proboscis-sheath of the fly and either penetrate the bite-puncture or pass directly through the skin.

The incubation period of the nodules in man is less than one year. Endemic regions are Central American Neotropica (vectors *Simulium metallicum*, *Simulium callidum*, and *Simulium ochraceum*) and Tropical Ethiopia (vector *Simulium damnosum*).

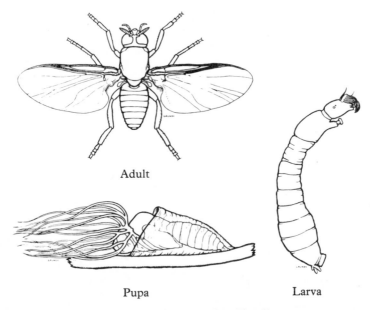

Adult

Pupa   Larva

Figure 31   Black-fly, *Simulium*

Recognition of Black-flies. *Simulium* (Fig. 31) are small (1 to 5 mm.) stoutly built, hump-backed flies with rather short (for

86

Nematocera) antennae of typically nine to eleven segments. These are sparsely haired or pilose in *both* sexes. The wings are broad and clear and often display an iridescent sheen. There are a few stout veins characteristically concentrated at the anterior edge of the wing. Male black-flies have their compound eyes touching (holoptic), while the eyes of the female are distinctly separated (dichoptic).

Habits of Black-flies. Only the females of *Simulium* snip the skin and suck blood, leaving a characteristic globule of coagulating blood at the site of the puncture. The flies may appear in such enormous numbers that they would certainly kill any available human beings who were unable to protect themselves by their clothing, by lighting smoky fires ('smudges'), or by retreating into dwellings. They attack in broad daylight and also occasionally by moonlight and are particularly active towards evening and during overcast thundery weather. *Simulium damnosum* in Africa bites chiefly in the morning and at dusk and rarely feeds above three feet from the ground. Although usually remaining near the breeding area, *Simulium* occasionally cover considerable distances, especially after rain has fallen.

Life-history of Black-flies. Several hundred dark coloured eggs, which are pale when first laid, are cemented to rocks or weeds in running streams. Some species attach their eggs to the submerged tips of plants. A high oxygen content such as is present in mountain-streams is essential to the immature stages. Five to twelve days are required before hatching. The larva resembles a tiny Indian-club in appearance. It anchors itself in an upright position to a stick or stone by means of a caudal sucker. There is a pair of prominent mouth-brushes, and a false-leg (with sucker) projects from the front of the thorax and is occasionally used as a foot. The larval food is composed of such organic particles as can be combed from the water. Larvae undergo six moults, and the pupal stage is reached after about three to ten weeks. The pupae are partially enclosed in a basket-like cocoon which is glued to the stream-bed or underwater objects. Respiration takes place through a pair of branched horns in contact with the running water (not atmospheric air). This stage requires approximately three to six days. Some species have five or six generations in a season. The egg or the

87

larval stage may survive over the winter period. The life-cycle varies according to conditions from about a fortnight to many months.

Control of Black-flies. The aquatic immature stages and the habits of the adult black-fly make control difficult. Natural methods such as the use of fish which prey on the larvae and parasitic protozoa have been used, but effective control relies upon the application of insecticides to the rivers and streams in which the larvae are living (eggs and pupae appear to be little affected by insecticides), control of the adult having little effect.

Suborder BRACHYCERA (*brachys*, short; *keras*, horn or antenna). These flies may be said to form the connecting link between the Nematocera with long antennae, and the Athericera with ultrashort antennae. They are of rather robust build and include the gadflies, which contain species of medical importance. A typical brachycerous antenna consists of some three segments terminating in an annulated style (representing fused and reduced antennal segments).

Gadflies (Tabanidae)

Female gadflies are ferocious biters and blood suckers, though the males, which are rarely seen, feed only on flower and plant juices. The distribution is world-wide, wherever suitable hosts, usually cattle and horses, are available. There is a tendency for seasonal occurrence. The adults suddenly appear and, after tormenting man and beast for a few weeks, as suddenly disappear. Their numbers are associated with the conditions in the breeding grounds during the previous year or two. Generally speaking abundant moisture is advantageous to the larvae and drought unfavourable.

Diseases caused and transmitted by Gadflies. The BITE causes immediate pain, but the saliva is believed to be non-toxic. Local swelling and ulceration may result, however. Large gadflies will, if permitted, feed for ten minutes and imbibe about 0·2 cc. of blood. Owing to the large size of the bite-puncture, the wound may bleed for a short time after the fly has ceased to feed.

Tularaemia. This is an infective plague-like disease of various small mammals, chiefly rodents. It is caused by certain bacilli which may be transmitted to man in the United States by *Chrysops discalis*. The transmission is mechanical and the mouth-parts of the fly may remain infective for about eight days.

Filariasis. *Loa loa* is the parasitic helminth causing 'eye-worm' infestation and 'Calabar swellings'. It is transmitted by *Chrysops dimidiata* and *Chrysops silacea* in the Tropical Ethiopian region. The microfilariae are deposited in the wake of the female worms and are imbibed by the vectors in a meal of peripheral blood during the day-time. After some ten days' development in the thoracic muscles and fat body, the mature larvae migrate to the blood space within the beak (labium) of the fly, which remains infective for about a week. On reaching man through the bite-puncture the worms disappear into the subcutaneous tissue and mature slowly. They may persist in the human body up to 15 years.

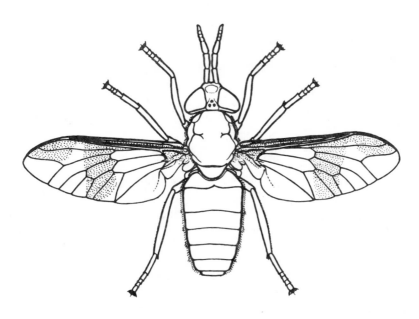

Figure 32   Deer fly, *Chrysops*

Recognition of Gadflies. There are well over 2,000 species of gadfly, all of which exhibit a characteristic stoutness of build and lack of bristles. They vary from species about the size of a housefly to those measuring two and a half inches across the outspread wings. The pair of compound eyes are often very large and prominent, and are invariably of most marked and beautiful iridescence. They occupy almost all of the head-mass in the males. The eyes of the females are separated (dichoptic), while those of the males touch (holoptic).

The wing pattern is diagnostic (Fig. 32). The costal vein extends around the entire margin. There are four rather angulated adjoining closed cells in the body of the wing. One of these, the hexagonal almost separate cell in the centre of the wing, is known as the 'discal cell'. There are a number of posterior cells widely open to the posterior margin. For reasons of medical convenience gadflies may be separated into two groups or subfamilies (Fig. 33).

(1) Those *without spurs* on the tibiae of the hind pair of legs TABANINAE, of nuisance value only.

(2) Those *with spurs* on the tibiae of the hind pairs of legs PANGONINAE, including *Chrysops*, of medical importance.

Both of these main groups or subfamilies may be further divided into two.

(1) Tabaninae (without spurs)

(a) With *mottled* wings...*Haematopota* or 'Clegs'.

(b) Usually with *clear* wings; and with a short spur on the upper side of the third antennal segment . . . *Tabanus* or 'horse flies'.

(2) Pangoninae (having spurs)

(a) With a very long proboscis or beak...*Pangonia* of doubtful nuisance value.

(b) With a medium length beak; and antennae of apparently seven segments (*i.e.*, three segments, with annulated style) projecting straight forward and longer than the head...*Chrysops* or 'deer flies' of medical importance.

Life-history of *Chrysops* 'deer-flies' (typical of other Gadflies). The female *Chrysops* lays several hundred eggs in a single layer in an oval-shaped patch on vegetation close to water (*Tabanus* generally lays its eggs several layers deep). The eggs are white when first laid but become darker after exposure. They hatch in approxi-

90

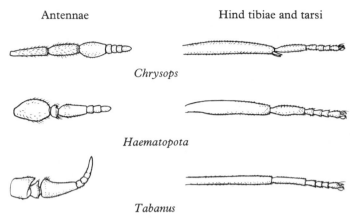

Antennae   Hind tibiae and tarsi

Chrysops

Haematopota

Tabanus

Figure 33   Tabanidae

mately five to seven days. The larvae are found in watery, muddy situations or in sodden vegetation. Some species are predaceous; others are saprophagous, *i.e.*, they eat rotten plants. They are elongate, taper at either end and are markedly segmented. The presence of Graber's organ, which is usually easy to see within the tenth and eleventh segments, is characteristic of gadfly larvae. Migration to drier ground takes place just before pupation, after the larvae have moulted six times. The larval stage may occupy several months to two years. The pupae are chrysalis-like. One to three weeks are required for the development and emergence of the adult *Chrysops*. The life-cycle may be completed in the year in which the eggs are laid, or delayed into the year following.

Control of Gadflies. Little work has been done on control by anti-larval measures owing to lack of knowledge of the habits of these flies. Measures designed to protect against the bites of adult flies must therefore be adopted. Such measures, which have been suggested as a protection against the vectors (*Chrysops silacea* and *Chrysops dimidiata*) of Loiasis, are:

1. Screening of houses. This has a limited effectiveness.
2. Repellents have been shown to give protection for two to three hours against the bites of *Chrysops* in the open.
3. The wearing of clothing which gives protection against bites.

It is doubtful whether insecticidal spraying directed against adult gadflies would be effective, owing to their habits. It is possible that the use of larvicides and natural control measures against larvae may be successful, but much more research into the bionomics of these flies is required.

Suborder ATHERICERA (*ather*, a spike; *keras*, horn or antenna). These flies are the most specialized and numerous of all Diptera. They are found in most parts of the world in an endless variety of roles. In size, those of medical importance range from the tiny Phorids and 'eye-flies' to the large 'blue-bottles', tsetse flies, and warble-flies. Athericera have a very short pair of antennae of three segments. The third and terminal segment is much the largest and there is always a projecting bristle or arista (representing three antennal segments), which is invariably dorsal in position. These flies are sometimes termed Cyclorrhapha, as the adult forms emerge from the puparium through a *circular* split in the chitinous case. The pupa is always enclosed in the unmoulted skin of the third stage larva. This enclosed pupa is termed a puparium. The larval forms have vestigial heads, usually tapering to a point. The mouth-parts are reduced to a pair of simple tearing hooks, and the posterior end is usually truncated for protection of the characteristic pair of posterior breathing spiracles.

Diseases caused by Athericerous Flies:
Nuisance value. Various species are sometimes present in such large numbers as to make eating and drinking out of doors very difficult. As many of the flies are attracted by sweat, sweet food, and, in some cases, by any moisture (such as drinking water, urine or faeces) they may show great persistence in alighting on the person, food, drink or excreta.

The bite. Only a few species of Athericera pierce the human skin and suck blood. Of these the common biting stable-fly, sometimes known as the biting house-fly (*Stomoxys calcitrans*), and the tsetse fly (*Glossina*) are the most important. *Stomoxys* rasps a puncture in the skin by means of the teeth at the extremity of its horny proboscis or beak. This causes a sharp pricking sensation which is very distressing to the host, if the flies attack in swarms. Secondary infection may occur after *Stomoxys* has flown off, as other non-biting flies may then contaminate the wound while feeding on the exudate. The tsetse fly (*Glossina*) may or may not cause a sharp pricking pain when attempting to feed. The after effects of the bite are not usually marked or prolonged. *Both* male and female biting Athericera pierce the skin and suck blood, whereas in those Nematocera and Brachycera which suck blood *only* the females are responsible, the males feeding on flower or fruit juices.

Myiasis. This is the invasion of the tissues of man by some stage, usually the larvae, of athericerous flies. It may be divided into three kinds:

(1) SPECIFIC MYIASIS, when the larvae develop only in (or in the case of the Congo floor-maggot, *Auchmeromyia luteola*, on) living flesh. Examples are *Chrysomyia bezziana* (India), and *Cordylobia anthropophaga* (Africa).

(2) SEMI-SPECIFIC MYIASIS, when the larvae invade human wounds or sores, though normally breeding in carrion or garbage. Examples are *Cochliomyia macellaria* (America), and *Lucilia sericata* (China).

(3) ACCIDENTAL MYIASIS, when any or all stages of the fly may be found in or passed from the human alimentary tract or uro-genital organs. This type of myiasis is usually acquired by ingesting eggs or larvae with food. It may cause severe digestive disturbance and other symptoms. Examples are *Fannia canicularis*, and *Megaselia scalaris*.

Control of Myiasis. Prevention must depend on such measures as the protection of wounds and sores against flies, good sanitation and the use of insecticides.

Diseases transmitted by Athericerous Flies:
Mechanical transmission of bacteria, protozoa, protozoal cysts, helminth ova, and other pathogenic organisms. Any flies which habitually settle on man's food or person after crawling on surfaces contaminated with pathogenic organisms (in latrines, on carcasses, sores, or sputum-deposits) may infect man with the organisms. This method of pathogenic transmission is threefold:

(1) By DIRECT TRANSPORTATION of the organisms on the sticky pads (pulvilli) on the extremity of the fly's legs, or on the body hairs. Houseflies and bluebottles are constantly brushing off particles attached to their hairs and bristles on to surfaces where they have alighted.

(2) By the VOMIT-DROP. Houseflies and bluebottle flies have a common habit of regurgitating the liquid contents of their crop and then partly or wholly reimbibing the bubble-like drop. Ring marks on windows left by flies are due to this discharge and intake. These vomits are laden with bacteria and easily contaminate food. This method of regurgitation is sometimes used by the fly partially to dissolve hard food such as crystals of sugar.

(3) By FAECAL DISCHARGE OR DEPOSIT on human food or tissues. Houseflies and bluebottles habitually excrete dark fluid spots which may be laden with pathogenic bacteria. Both the larvae and adults of houseflies may ingest the eggs of *Ascaris* worms, also the spores of anthrax and tetanus, and pass them in the faeces of the adult fly. The larva may retain these organisms through the pupal to the adult stage before voiding them. Epidemic typhoid, epidemic diarrhoea, bacillary dysentery, amoebic dysentery, cholera, tuberculosis, catarrhal conjunctivitis or 'pink eye' and ophthalmia are some of the 30 possible diseases conveyed by athericerous flies.

African trypanosomiasis or sleeping sickness. This is caused by certain trypanosomes which are transmitted to man by some ten species of tsetse flies or *Glossina*. These flies are found in Tropical Ethiopia in areas specially suited to their requirements, known as 'fly-belts'. Game animals form the chief endemic reservoir but infected natives are also a focus of the disease. Direct mechanical transmission may occur when a tsetse fly, interrupted while feeding on an infected animal, flies within two or three hours to a human being and inserts its infective mouthparts in order to continue feeding. The usual method of transmission, however, is through the bite of the fly, after the previously ingested trypanosomes have undergone a development phase usually lasting about 20 days. page 116).

It is always helpful when a large number of species has to be studied, to divide them into groups. As the scientific classification is rather complicated, the Athericera of medical importance are here arranged as practically as possible into three simple groups based on their habit-relationship to man:

(1) 'INDOOR FLIES' including the common house-fly (*Musca domestica*), the lesser house-fly (*Fannia canicularis*), the latrine fly (*Fannia scalaris*), and the Congo floor-maggot fly (*Auchmeromyia luteola*).

(2) 'SEMI-OUTDOOR FLIES', including the 'bluebottle' flies (*Calliphora* and *Cochliomyia*, also *Chrysomyia*), the 'greenbottle' flies (*Lucilia*), the 'blackbottle' flies (*Phormia*), the tumbu fly (*Cordylobia anthropophaga*), the non-biting stable fly (*Muscina stabulans*), the biting stable fly (*Stomoxys calcitrans*), the Phorids, the fruit flies (*Drosophila*), the 'eye-flies' (*Hippelates* and *Siphunculina*), the Sepsids, and the 'cheese-skipper' fly (*Piophila casei*).

(3) 'OUTDOOR FLIES', including the 'grey-bottle' or flesh-flies (*Sarcophaga* and *Wohlfahrtia*), the tsetse flies (*Glossina*), and the bot-flies (*Gasterophilus, Oestrus, Dermatobia, Rhinoestrus, Hypoderma*).

1. Indoor Flies

THE COMMON HOUSE-FLY (*Musca domestica*)·(Fig. 34) This fly has an almost world-wide distribution wherever there are human dwellings.

Diseases associated with the House-fly. It is perhaps the most notorious mechanical vector of pathogenic organisms, such as cause typhoid and dysentery, to human food. Dysentery bacilli remain alive in the gut of the fly for more than four days and the cysts of *Entamoeba histolytica* for two days. House-fly larvae may also on rare occasions cause semi-specific and accidental myiasis.

Recognition of the House-fly. All *Musca* have the aristae plumed on both sides. The wing is diagnostic in having the fourth vein bent gently up almost to meet the third vein just above the tip of the wing. As there are several species of *Musca* concerned with the transmission of epidemic diarrhoea and bacillary dysentery in

95

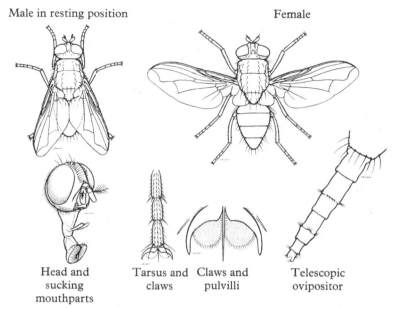

Male in resting position   Female

Head and   Tarsus and   Claws and   Telescopic
sucking   claws   pulvilli   ovipositor
mouthparts

Figure 34   House-fly, *Musca domestica*

different parts of the world, it is not always easy to determine the exact species. *Musca domestica* is about a quarter of an inch long and has four longitudinal dark stripes on the dorsum of the thorax. The female abdomen is mostly grey in colour whereas the male displays a noticeable amount of tawny-yellow on this part of the body. The wings usually diverge at about 60° in the resting position.

Habits of the House-fly. The adults are fond of feeding and breeding on human faeces and also of visiting human food, particularly if it is sweet. The size of the pathogens ingested is usually microscopic, owing to the small dimensions of the grooves in the sucking lips of the retractable mouth-tube. House-flies are capable of flying several miles from their breeding ground.

Life-history of the House-fly. Several hundred elongate whitish eggs are laid in crevices of fermenting organic matter, such as horse manure, garbage and faeces, by means of a telescopic ovipositor situated at the end of the female abdomen. They may be deposited in batches of about a hundred eggs and are small, white and elongately oval and visible to the naked eye. In warm weather the larvae may hatch in about eight hours. They are then 2 mm. long. The larvae feed ravenously on the available food supply and may pupate in three days to a week, if food and temperature are suitable. When about to pupate the blind legless white maggot crawls to a drier situation, where it shrinks and hardens into a brown barrel-shaped puparium. This is about a quarter of an inch long and may be found just below the surface, or several feet deep in the earth adjacent to the larval food supply. The adults may emerge from the pupa in three days, or the winter months may be passed in this stage. They are unable to breed for several days after emergence. The life-cycle may be completed in a minimum time of five days, but usually occupies two or three weeks.

THE LESSER HOUSE-FLY (*Fannia canicularis*) (Fig. 35). This is a common house-fly of temperate climates.

Diseases associated with the Lesser House-fly. It is a mechanical vector of pathogenic organisms, besides causing accidental myiasis of the intestines.

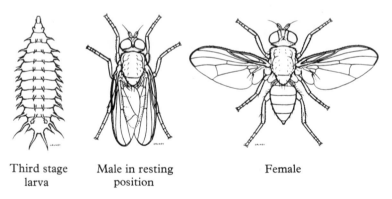

| Third stage
larva | Male in resting
position | Female |

Figure 35   Lesser House-fly, *Fannia canicularis*

Recognition of the Lesser House-fly. The genus *Fannia* always has bare aristae. The wing venation characteristically shows the fourth vein extending straight to the wing margin, well below the wing-tip and almost parallel to the third vein, which ends

97

exactly at the wing tip. The sixth vein is remarkable for its shortness. There are, however, a great many outdoor-flies with this type of wing venation. The lesser house-fly *Fannia canicularis* is about half the size of the common house-fly. The thorax has three longitudinal dark stripes on its dorsal surface. It typically folds its wings more over its abdomen than *Musca domestica*. The male has the sides of the second and third abdominal segments noticeably yellow and translucent. The female has a more heart-shaped and plumper abdomen than the male and the colour is predominantly grey.

Habits of the Lesser House-fly. The males often visit dwellings in large numbers during the warm months of the year. They play continuously in the air, between periods when they settle on hangings and lampshades, which they soil with their excreta. The females are more plentiful near outdoor garbage and especially near earth soiled with urine.

Life-history of the Lesser House-fly. This is somewhat similar to *Musca domestica*. The eggs hatch about 24 hours after being laid in fermenting vegetation, manure, night soil or semi-liquid faeces. The larvae require about one week to a month to reach the pupal stage. They have a very characteristic appearance, for they are flattened with many spine-like fleshy processes. These enable the larva to position itself in semi-liquid materials. The pupa usually remains about one week before the adult fly emerges. It resembles a dried up or dead larva since it retains the final larval skin together with its projecting processes.

THE LATRINE FLY (*Fannia scalaris*). This is a fly of rather similar appearance and distribution to the lesser house-fly. It causes and conveys the same diseases. It is, however, darker in colour than *Fannia canicularis* and bears a distinct tubercle on the tibia of the middle pair of legs. The larva is even more characteristically decorated with fleshy spine-like protrusions, since in the latrine fly maggot these are 'feathered'. It breeds almost entirely in faeces or soiled earth.

Control of Flies. *Musca domestica.* Control measures may be summarized as follows:

1. Control of breeding places.
2. The use of insecticides.
3. The protection of food.

Control of breeding places. This is the all-important measure of fly-control for which there can be no substitute. It is a matter of good sanitation and includes the proper disposal of waste matter, denial to flies of access to breeding-grounds by such measures as the fly-proofing of latrines and rendering breeding material unsuitable, for example, by tight-packing of horse manure.

The use of insecticides on breeding-places is a secondary measure, which should only be used when control is urgently required or when breeding-places cannot be dealt with in any other way. The object is to kill flies alighting on the surface or emerging from pupae.

The use of insecticides. As there are many loopholes in the control of flies by prevention of breeding, control by insecticides is very important. The insecticides used against flies are divided into two groups:

1. Those which have a residual effect but which kill relatively slowly—'residual sprays'.

2. Those which cause a quick 'knock-down' of insects but which do not persist—'space sprays'.

Insecticides of the first group are the mainstay of control; those which cause a quick knock-down have a more limited value. Residual insecticides may be in the form of wettable (water dispersible) powders used for spraying porous surfaces, and emulsion concentrates for smooth non-absorbent surfaces.

Aerosol dispensers are a convenient form of apparatus for dispensing knock-down sprays. Concentrated insecticide is dissolved in an inert, liquified gas which is retained under pressure in a 'beer can' type of container. The gas employed may be 'Freon' (di-chloro-di-fluoro-methane) or a mixture of hydrocarbons; when the pressure is released by opening a nozzle on the container, the liquid gas escapes, leaving fine particles of insecticide suspended in the air.

Where flies are present in large numbers, all indoor surfaces must be treated with a residual insecticide at the intervals given on

Fannia canicularis (Linnaeus, 1761)

Stomoxys calcitrans (Linnaeus, 1758)

Musca domestica (Linnaeus, 1758)

Glossina palpalis (Robineau-Desvoidy, 1830)

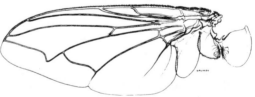

Calliphora erythrocephala (Meigen, 1826)

Note.—These species cannot be identified by their wings alone

Figure 36   Wings of some Athericerous flies of medical importance

the container. Spraying from the air is effective against adult flies but the effect is short-lasting.

Selective spraying is used if flies are not so numerous as to be beyond control except by universal indoor spraying. The purpose of selective spraying is to kill the maximum number of flies with the minimum expenditure of spray. When selective spraying is being carried out, the following areas should be treated:

1. Walls and ceilings of kitchens, stores, preparation rooms, butcher's shops, slaughter-houses and dining-rooms.

2. Indoor resting-places of flies; corners of walls near ceilings, electric-light cords, lampshades, windows and surrounds (inside and outside).

3. Outdoor resting-places; swill-bins, incinerators, latrine-seats, and screens, stables, byres, pigsties and so forth.

Where residual spraying has been carried out, the use of aerosols should not be necessary. The main requirement for this type of spray is against mosquitoes; it is also effective against flies.

When breeding-places are being treated with residual insecticides, the ground for a distance of six feet round the site should also be treated to destroy any young flies which may emerge there. This method is not of value for treatment of breeding-places to which fresh material is continually being added or for areas in which new breeding-places appear daily.

Protection of food. The final defence against fly-borne diseases is the protection of food by fly-proofing. Food should be protected at all stages; if possible it should always be transported in fly-proofed containers and be stored, prepared and eaten in fly-proofed rooms.

Food liable to contamination by flies should always be kept covered up and preferably in a fly-proofed safe. Eating and cooking utensils should also be carefully protected.

Protection of food is the most important single means of guarding against fly-borne diseases because it is not possible to control breeding places.

Buildings are most effectively fly-proofed by covering the windows with fly-wire, and by making self-closing doors of fly-wire on light wooden frames. If fly-wire is not available old mosquito or sandfly-netting is satisfactory, though less durable. Doorways can be covered with fish-netting hung loosely as a curtain that moves gently in the wind.

Another important measure is to keep all latrines fly-proofed so that flies are denied all access to pathogenic organisms; this is particularly important in countries where flies are prevalent and dysentery and diarrhoea are common.

Repellents. Repellents are effective for 2 to 3 hours. Care should be taken to keep it out of the eyes, and away from plastic substances such as fountain-pens or watch glasses, upon which it may exert a solvent action.

The Lesser House-fly and Non-biting Stable Fly. Control is based upon the same general principles used in the control of *Musca domestica*. Particular attention should be paid to the types of breeding place favoured by these flies.

THE CONGO FLOOR-MAGGOT FLY (*Auchmeromyia luteola*). This is very widely distributed in the Tropical Ethiopian region.

Disease caused by Congo floor-maggots. The larvae issue forth at night from the dust or from crevices in the floor of native huts and suck the blood of persons sleeping on the ground. If the larvae are numerous this may result in the host becoming anaemic.

Recognition of the Congo Floor-maggot Fly. The adult flies are a brownish-yellow colour, and are nearly related in size, build and wing venation to the common bluebottle. The abdomen, however, is noticeably more elongate and pointed. The extra length of the second abdominal segment when compared to the remaining visible segments differentiates it from the tumbu fly.

Habits of the Congo Floor-maggot Fly. The adults are never found far from dwellings, and spend much of their time resting under the shelter of the walls and thatch of native huts. They seem to dislike excessive heat and fly chiefly in the early morning or after sunset. Sweetened food, fallen fruit and fresh excreta are attractive to them.

Life-history of the Congo Floor-maggot Fly. The eggs are laid

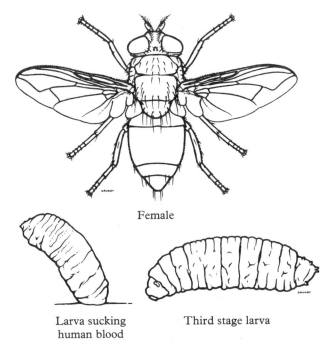

Female

Larva sucking
human blood

Third stage larva

Figure 37   Congo floor-maggot fly, *Auchmeromyia luteola*

indoors in soil or matting which is tainted with sweat or urine.
They hatch into larvae in about two days. The larvae, especially in
the early stages, are somewhat similar in appearance to bluebottle
maggots or 'gentles'; they are differentiated by having five pairs
of fleshy posterior processes as well as characteristic posterior
spiracles. They may pupate within a fortnight if frequent blood
meals are available, or be delayed by partial starvation for three
months. The pupa is brown and barrel-shaped and found in dry
dust or crevices with the larvae. The adults emerge in less than a
fortnight.

Control of the Floor-maggot Fly. Control depends upon
cleanliness and good sanitation. The use of high beds gives protection, while infected huts can be rendered habitable by firing the
ground or by digging up floors and removing the surface soil,
which can then be disinfected. Sleeping-mats and blankets should

103

be searched for larvae as it is possible that they may be transported in such impedimenta.

2. Semi-outdoor Flies

BLUEBOTTLE FLIES (*Calliphora*). These are the blow-flies of the temperate regions.

Diseases caused and transmitted by Bluebottles. The larvae may cause semi-specific and accidental myiasis. The adults are notorious mechanical vectors of pathogenic organisms.

Bluebottle fly, *Calliphora*, female

Figure 38

Recognition of Bluebottle Flies (Fig. 38). They are large (unless the larvae were underfed) burly rather bristly flies of a dusty metallic-blue colour. The aristae are plumed both sides and the cheeks are very hairy. The wing venation somewhat resembles *Musca* but the fourth vein bends up with a sharpish angle, not gently, almost (there is a small gap) to meet the third vein above the wing-tip.

Habits of Bluebottle Flies. They are very strong fliers and are able to find meat and carcasses from considerable distances. They are fond of feeding on liquid from faeces as well as visiting human food supplies.

104

Life-history of Bluebottle Flies (Fig. 39). Several hundred elongate white eggs are laid in closely-packed batches in crevices of fresh meat or putrid carcasses, by means of the telescopic terminal ovipositor of the female (Fig. 39). The larvae, which hatch out in summer weather in about eight hours, feed voraciously in the liquid liberated in the food supply by their pair of tearing mouth-hooks. Under suitable conditions the third stage larva, after feeding for about nine days, will migrate from the food supply at night, and crawl away to pupate some three to nine inches below the soil surface. The adult fly emerges from the red-brown puparium in about a week. By means of an eversible bladder on the head, newly

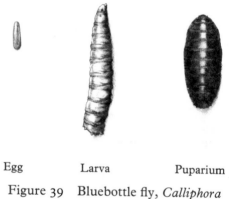

Egg   Larva   Puparium

Figure 39   Bluebottle fly, *Calliphora*

emerged bluebottles, house-flies and their near relatives are able to emerge through a considerable depth of loose earth or debris. The same organ (ptilinum) enables the adult fly to push open the pupal-cap. The life-cycle is usually about three weeks and there are some four generations in a year.

TROPICAL BLUEBOTTLE OR SCREW-WORM FLIES (*Chrysomya* and *Cochliomyia* [= *Callitroga*]). These flies replace *Calliphora* in many of the warmer parts of the world.

Diseases caused and transmitted by Screw-worm Flies. SPECIFIC and SEMI-SPECIFIC MYIASIS. The larvae of the primary screw-worm fly (*Cochliomyia hominivorax*) are found in the United States Nearctic, Central American, Andean and Brazilian Neotropic regions, and infest living tissue, occasionally entering through the unbroken skin. Although one or two larvae cannot break down tissue faster than the natural healing process, yet a number of the larvae working together may cause dangerous necrosis of flesh, brain or bone. The mortality in cases of nasal and aural myiasis is estimated at approximately 8%. The common or New-world screw-worm fly (*Cochliomyia macellaria*) is often attracted by the foul lesions caused by *Cochliomyia hominivorax* so that its larvae are sometimes found in the same wounds. By themselves the larvae of *macellaria* tend to clean the lesion of dead tissue rather than cause necrosis, but in the presence of the larvae of *hominivorax* they add to the general destruction. The Old-world screw-worm fly (*Chrysomyia bezziana*) found in the Ethiopian, Oriental and Australian regions causes a similar specific myiasis to *Cochliomyia hominivorax*, which often results in serious and disfiguring necrosis of tissue and bone. Ophthalmic infestations are especially dangerous. The adult flies of several species of tropical bluebottle or screw-worm flies are mechanical vectors of pathogenic organisms.

Recognition of Screw-worm Flies. They are mainly burnished purple-blue colour, often decorated with dark stripes on the thorax and black bands on the abdomen. They may be distinguished from *Calliphora* and *Lucilia* by the absence of bristles on the dorsum of the thorax. The antennae are usually orange and the aristae are plumed on both sides. The wing venation is similar to *Calliphora*.

Habits of Screw-worm Flies. They are similar to *Calliphora* bluebottles in their breeding habits and fondness of feeding on faeces and such human food as meat and bazaar sweets.

Life-history of Screw-worm Flies. *Cochliomyia hominivorax*, a good example of the genus, may deposit up to 300 eggs in five minutes on or near a wound or discharging aperture. The young screw-worms or maggots burrow deeply for a period of from four days to three weeks, and when fully grown ($\frac{2}{3}$-in. long) they leave the host's body and pupate in the ground. The adult fly appears in as little as one week or as long as several months. *Cochliomyia* *macellaria* normally breeds in carcasses but readily deposits its eggs in diseased tissue. The life history of most species of screw-worm flies is similar to that of the temperate region blowflies and garbage flies.

GREEN-BOTTLE FLIES (*Lucilia*). These are temperate and tropical-zone blowflies.

Diseases caused and transmitted by Green-bottle Flies. SEMI-SPECIFIC MYIASIS. *Lucilia* normally breeds in carcasses, but certain species have a tendency to deposit their eggs in wounds and sores, the larvae not only devouring the diseased tissue but invading the living flesh. This occurs chiefly in the Chinese Orient and the southern Mediterranean Palaearctic region. These flies are also concerned in ACCIDENTAL MYIASIS and the MECHANICAL TRANSMISSION OF PATHOGENIC ORGANISMS.

Recognition of Green-bottle Flies. They are medium-sized flies, usually smaller than *Calliphora* and of a beautiful metallic or burnished green or blue colour. There are no markings on the thorax, but there are two longitudinal rows of thoracic bristles on the dorsum. The aristae are plumed on both sides and the wing venation is similar to *Calliphora*.

Habits of Green-bottle Flies. These are similar to *Calliphora*, though *Lucilia* adults are perhaps even more fond of feeding on faeces and putrid matter.

Life-history of Green-bottle Flies. Very similar to that of *Calliphora*.

BLACK-BOTTLE FLIES (*Phormia*). These flies might be described as being midway in relationship between the screw-worm flies (*Cochliomyia* and *Chrysomyia*) on the one side, and the green-bottle flies (*Lucilia*) on the other. They may be of a metallic black, blue, or green colour. Some species are MECHANICAL VECTORS OF PATHOGENIC ORGANISMS. The larvae of the black blow-fly (*Phormia regina*) sometimes cause SEMI-SPECIFIC MYIASIS.

Control of Blowflies. Control must primarily be based upon good sanitation and the protection of food, especially meat, from the attentions of the fly. Residual insecticides are effective against adults and may be applied to breeding-places to kill emerging flies. Insecticides are best applied in the form of dusts to kill ovipositing flies. Dusting may have to be repeated weekly or fortnightly to cure bad infestations, in situations where the ordinary sanitary measures cannot be applied. Residual sprays may also be used.

THE TUMBU FLY (*Cordylobia anthropophaga*). This large yellow-brown fly is well known in the Tropical Ethiopean region, to which territory it is limited.

Diseases caused by Tumbu Fly maggots. SPECIFIC MYIASIS. The first stage larva attaches to the skin and wriggles into the nearest wrinkle. With the aid of its mouth-parts it rapidly cuts an entrance (in from half a minute to several minutes), and penetrates just under the skin. There may be little reaction or there may be intense itching and white blebs on the skin. The later development

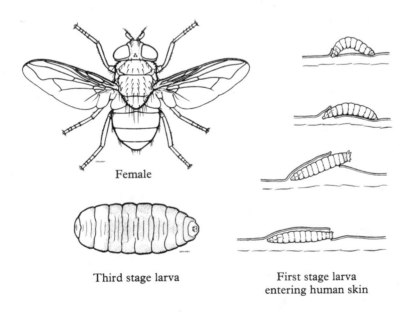

Female

Third stage larva

First stage larva
entering human skin

Figure 40   Tumbu fly, *Cordylobia anthropophaga*

of the larva produces a boil-like swelling. Sloughing and gangrene may occur if a number of larvae are localized in one area.

Recognition of the Tumbu Fly (Fig. 40). This is a large brownish-yellow fly nearly related in size and structure to the common bluebottle fly. The aristae are plumed on both sides and the wing venation is similar to *Calliphora*. Tumbu flies are easily differentiated from Congo floor-maggot flies by their more rounded abdomen (*Auchmeromyia* is relatively elongate and pointed), as the four visible segments of the tumbu are almost equal in length (*Auchmeromyia* has the second segment very long especially in the female). They are, however, difficult to differentiate from some other African yellow-bottle flies not of medical importance.

Habits of the Tumbu Fly. The adult flies live in the vicinity of native villages, and are said to follow their chief host, the rat, into dwellings when the rainy season drives the rodents from their outdoor holes. They may be found in huts and houses all the year round, however, as they do not like intense heat and are very fond of sweet foods, fruit juices or blood.

Life-history of the Tumbu Fly. The females, which are capable of laying 500 eggs, seek damp, not wet, sandy soil contaminated by urine, perspiration or other excreta.

They dig small cavities for the eggs, which are laid a few in each hole, usually during two periods of oviposition of a hundred or so eggs in each. The eggs hatch into larvae in 24 hours to three days. These, *in situ*, may survive ten days or more of starvation. The young maggots are very sensitive to the presence of an animal body, and quickly seize and penetrate the skin of man, rat, or dog, if such a host comes near enough to be parasitized. The feet, forearms, scrotum and back are usual places of infestation in man. After nine or ten days' development in the living tissues, the mature larvae drop to the ground, and soon harden into puparia. The pupal stage takes about three weeks or more. The larva is able to live some 9 to 15 days without food at room temperature (68°F), but the life-cycle under favourable conditions takes a little more than a month.

Control of Tumbu fly. Control must be based upon good sanitation, avoidance of areas likely to be infested and of sleeping on the ground. The numbers of rats and dogs must be kept low. Little appears to be known about the effect of insecticides and repellents.

NON-BITING STABLE FLIES (*Muscina*). These flies are widely distributed in temperate zones.

Diseases caused and transmitted by Non-biting Stable flies. ACCIDENTAL MYIASIS has been recorded many times, accompanied by convulsions, general debility and nausea. The adults are mechanical vectors of pathogenic organisms.

Recognition of Non-biting Stable Flies. The grey-coloured adult flies are somewhat larger and stouter than the common house-fly, which they much resemble in general appearance. The aristae are plumed on both sides. The wing venation (in association with the non-biting mouth-parts) is diagnostic—the fourth vein bending very gently towards the third vein and terminating at the tip of the wing, leaving the cell between them open. In the biting stable fly (*Stomoxys*) the fourth vein is relatively more curved. Further, this fly (*Stomoxys*) always has a biting proboscis and the compound eyes when viewed laterally are kidney-shaped.

Habits of Non-biting Stable Flies. These flies may sometimes occur in such numbers as to become real pests. They readily enter houses in hot weather to lay eggs on vegetables which are beginning to decompose. They are less active and more clumsy in movement than common house-flies (*Musca domestica*).

Life-history of Non-biting Stable Flies. This is somewhat similar to that of *Musca*. Animal excreta and garbage form the chief breeding material.

THE BITING STABLE FLY OR BITING HOUSE FLY (*Stomoxys*). These flies have an almost world-wide distribution in temperate and tropical zones.

Diseases caused and transmitted by Biting Stable Flies. The BITES have a nuisance value and allow of secondary infection. AFRICAN TRYPANOSOMIASIS can be conveyed mechanically by the bite of *Stomoxys*.

Recognition of Biting Stable Flies. Though resembling the common house-fly (*Musca domestica*) in size, grey colour and general appearance, *Stomoxys* is easily differentiated from it and other flies by the presence of a long shiny brown biting proboscis or beak hinged below the head; by the arista being plumed on the upper side only; and by the wing venation (which is somewhat similar to *Muscina*), the fourth vein curving towards the third vein with a definite bend and ending below the tip of the wing. In both *Stomoxys* and *Muscina* the cell between the third and fourth veins is widely open at the wing margin. A further characteristic of *Stomoxys* is that the eyes when viewed from the side appear reniform or kidney-shaped.

Habits of Biting Stable Flies. Both the males and females attack cattle, horses and man in the open air and inside buildings. They suck up blood which oozes from the somewhat ragged bite-puncture. The proboscis is not deeply inserted as it is in the case of mosquitoes and tsetse flies. *Stomoxys calcitrans* generally rests on a wall with its head upwards. *Musca domestica* prefers to have the head downwards.

Life-history of Biting Stable Flies. The eggs are laid outdoors in stable manure or rotting vegetation. They are laid in scattered or irregular heaps of fifty or more. The life-history is very similar to that of *Musca* and occupies on an average from 25 to 37 days.

PHORID FLIES (*Phoridae*). These tiny flies belong to a family of world-wide distribution.

Diseases caused and transmitted by Phorids. ACCIDENTAL MYIASIS: the larvae may be ingested in food and reappear in the faeces. The adult flies are suspected of conveying pathogenic organisms to human food. *Megaselia scalaris* (= *Aphiochaeta scalaris*), which is widely distributed in the warmer parts of the world, has actually been known to occur in all stages within the lumen of the human gut.

Recognition of Phoridae. The flies are very small. The antenna has one large spherical segment which conceals the others. It bears a characteristic pubescent elongate arista. The wings when present show the anterior veins characteristically heavy in build and clustered on the front margin.

Habits of Phorids. They are active runners and have a curious humped-backed appearance. They are usually found among decaying vegetation or faeces.

Life-history of Phorids. They breed in partly empty milk bottles, faeces, and garbage filth. The larvae are dirty white and about 5 mm. long. They progress slowly in the manner of a 'looper' caterpillar. There are small tubercles and posterior spiracles raised on posterior chitinous processes. The pupae display a pair of characteristic horns. The life history takes about three weeks to a month.

FRUIT FLIES (*Drosophilidae*). These very small flies have a world-wide distribution in temperate and tropical zones.

Disease caused by Fruit-flies. ACCIDENTAL MYIASIS may occur through the ingestion of fruit containing eggs, larvae or pupae. Digestive disturbance may result. The adult flies are suspected of conveying pathogenic organisms to human food.

Recognition of Fruit-flies. These little flies are usually of a brown colour, and exhibit a swollen appearance. The aristae are usually feathered and the eyes are generally red. The adults have a habit of flying slowly or hovering in the vicinity of ripe fruit in houses. The wing venation is not unlike a simplified form of *Fannia*.

Habits of Fruit-flies. The adults are much attracted to rotting fruit, faeces, and vinegar vats. Outbreaks of fruit-flies often occur in catering establishments owing to the presence of uncleared garbage.

Life-history of Fruit-flies. The life-cycle occupies approximately a fortnight in warm weather. Even a single rotting banana will produce a great number of flies.

EYE-FLIES (*Oscinidae*). These small flies are limited to the tropical zone.

Diseases caused or transmitted by Eye-flies. NUISANCE VALUE; these flies are often great pests from their habit of hovering in front of the eyes, apparently attracted by the optic secretions. EPIDEMIC CONJUNCTIVITIS is transmitted mechanically by the black eye-gnat *Hippelates pusio*, of America. The Asian eye-fly *Siphunculina funicola* also transmits pathogenic organisms to eyes and wounds.

Recognition of Eye-flies. These are small, very shiny flies. The aristae are bare, the wing venation resembles a simplified form of *Fannia*, and there are never markings as in *Sepsis*. *Siphunculina funicola* is dark brown with the upper surfaces darkening almost to blue.

Habits of Eye-flies. Some species gather in swarms in buildings or on verandahs. They may be seen hanging on cobwebs, roof pendants etc. They display fondness for the secretions of the eyes, open sores, pus, and blood.

Life-history of Eye-flies. The larvae may feed on grass stems, roots and other vegetation or amongst decaying matter in the soil.

SEPSID FLIES (*Sepsidae*). These tiny flies are world-wide in their distribution.

Diseases caused or transmitted by Sepsids. ACCIDENTAL MYIASIS. The larvae have been recovered from the intestine and the uro-genital system. The adults are suspected of conveying pathogenic organisms to human food.

Recognition of Sepsids. The adult flies are rather pretty delicate flies with a globular head and a general blackish or reddish body-colour. The abdomen is invariably constricted at the waist and the general appearance is ant-like. They are usually shiny and the wings often bear a characteristic stigma-like spot.

Habits of Sepsids. The adults have a curious habit of running or standing and at the same time twisting their wings in the air. They are attracted towards excrement, and decomposing organic matter.

Life-history of Sepsids. The larvae are elongate and maggot-like. They are found in decomposing organic material, and occasionally hop after the manner of the well-known cheese skipper (*Piophila casei*). The posterior spiracles are raised on tubercles.

HOVER-FLIES (*Syrphidae*). The distribution of these medium to large, often brightly-coloured flies is world wide.

Diseases caused by Hover-fly larvae. ACCIDENTAL MYIASIS; occasionally the larvae of the rat-tailed maggot fly (*Eristalis tenax*) and of some species of *Helophilus* have been recovered from the human intestine. They have not caused any serious reaction.

Recognition of Hover-flies. The wing venation, which displays a 'false vein' as well as a tendency towards a second margin internal to the hind edge, is diagnostic.

Life-history of Hover-flies. While the rat-tailed maggot breeds in stagnant water contaminated with decaying organic matter, the larvae of many other species live on aphids infesting plants. They may be ingested with garden fruit or salads.

THE CHEESE SKIPPER FLY (*Piophila casei*). These small flies are of world-wide distribution in stored animal-protein food (not starchy products), such as cheese, bacon, and dried fish.

Disease caused by Cheese skipper maggots. ACCIDENTAL MYIASIS; the fully grown larvae or 'cheese skippers' are not uncommonly ingested in foods such as the highly-flavoured cheeses. They are very resistant to the human digestive juices and when swallowed in numbers may cause scarification of the gut. There is one record of invasion of the nasal cavity.

Recognition of Cheese Skipper Flies and their larvae. The adult flies are small (2–4 mm.) and shiny, with a rounded head and bare aristae. They are of stouter build and without the marked waist or wing spots of the Sepsidae. The mature larvae are elongate maggots (about 10 mm.) which have the remarkable power of bending the head segments until they can seize their posterior end. When the strain of holding is suddenly released, the 'cheese skipper' hops about a foot.

Habits of Cheese Skipper Flies. They run actively and are of very quick flight.

Life-history of Cheese Skipper Flies. The females lay their eggs in stored animal-proteins. The larvae resemble long, thin maggots. The pupae are sometimes mistaken for elongate seeds stuck in the wrappings of cheese and bacon or in the cracks of storehouse floors.

The life-cycle occupies about 12 to 14 days and the insects may multiply enormously when left undisturbed.

3. Outdoor Flies

TSETSE FLIES (*Glossina*). These are inhabitants of the Tropical Ethiopian region, where the different species restrict themselves to areas known as 'fly-belts' which appear to be suited to their specific habits.

Diseases transmitted by Tsetse Flies. There are some ten species concerned with the transmission of AFRICAN TRYPANO-SOMIASIS or Sleeping Sickness (Fig. 41). Of these species four are more important than the remainder. These are *Glossina palpalis, Glossina tachinoides, Glossina morsitans* and *Glossina swynnertoni.*

Recognition of Tsetse Flies. These flies (Fig. 42) are always of a brown colour. They may be smaller than the common house-fly, as is *Glossina tachinoides,* or about the size of a house-fly, as are *Glossina palpalis, Glossina swynnertoni* and *Glossina morsitans*; or larger than the common bluebottle fly, as is *Glossina fusca.* The aristae are diagnostically plumed on the upper side only, with the plumes repluned or feathered on either side (Fig. 43) (in *Stomoxys* the plumes are unbranched). The palps or sheaths of the mouth parts always project in front of the head, even when the cutting and sucking parts are hinged downwards to partake of a blood-meal. The wing venation is diagnostic in that the fourth vein has a particular forward curve which causes the 'discal cell' (space between the fifth and fourth veins) to resemble the outline of the blade of a 'butcher's cleaver'. The fourth vein approaches the third well above the wing tip.

Habits of Tsetse Flies. The different species show preferences in the choice of their local habitat.

Glossina palpalis (found chiefly in West Africa and the Congo Basin) is found mainly in well-wooded country near rivers or lakes. Crocodiles and other reptiles form important hosts.

Glossina morsitans (chiefly in eastern Central Africa) is mainly found in more open forest land, where mammalian game is plentiful.

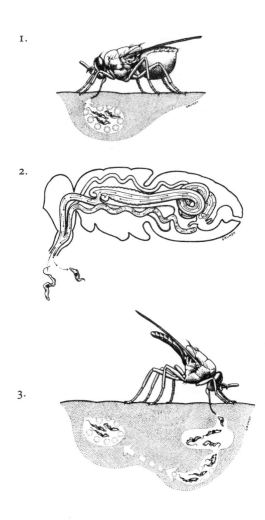

1. *Glossina* ingesting trypanosomes in blood of human host.

2. Cyclic development of trypanosomes within gut, peritrophic space and salivary glands of fly. This takes about twenty days.

3. Injection of metacyclic trypanosomes (infective stage) into blood stream of new human host during subsequent blood meal.

Figure 41   Trypansomiasis cycle in man and the *Glossina* tsetse fly

116

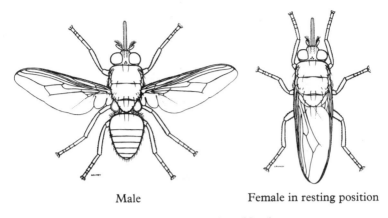

Male   Female in resting position

Figure 42   Tsetse fly, *Glossina*

Glossina swynnertoni prefers even more open, drier country where there is an abundance of game animals. *Glossina tachinoides* is able to flourish in areas where man is the principal or only host.

Both male and female tsetse flies bite and suck blood. They typically attack in the day time and are attracted by moving objects of a dark colour.

Life-history of Tsetse Flies. The females are remarkable for *not* depositing eggs or first-stage larvae, as they rear one larva at a time

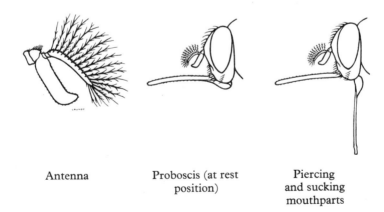

Antenna   Proboscis (at rest   Piercing
                 position)   and sucking
                                             mouthparts

Figure 43   Tsetse fly, *Glossina*

117

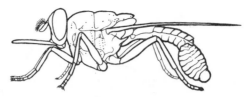

Female depositing third stage larva

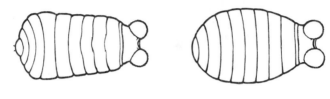

Larva   Pupa

Figure 44   Tsetse fly, *Glossina*

to the third stage within the abdomen (Fig. 44). When this reaches maturity, after moulting twice inside the parent, it is passed out in a carefully selected shaded spot so that it may pupate. This larvipositing of mature maggots takes place once every 10 or 12 days during the breeding period of the parent fly. The larva crawls under cover of the loose sand or decaying bark on which it is deposited and hardens into a puparium. The adult form emerges from the pupal stage in about three or four weeks. Both larvae and pupae may be differentiated from those of other flies by the characteristically large pair of dark chitionous knobs at their posterior extremity. (These knobs protect the pair of posterior spiracles.)

Control of Tsetse Flies. This is attempted by measures designed to alter unfavourably the habitat of the fly.

Insecticides. These are used, but with limited success. Residual insecticides are effective against adult tsetse flies, but it is difficult to devise the best means for bringing the relatively few and scattered flies into contact with the insecticide and to get effective penetration of dense bush. The best results have been obtained with smokes put down by aircraft in open country against *Glossina morsitans* and *Glossina swynnertoni*, but smokes are ineffective

118

against *Glossina palpalis* because they do not penetrate the thick bush in which it lives sufficiently to give adequate kills. Smokes can also be put down by ground generators.

Fine sprays of insecticide are effective when delivered from aircraft, in droplets ranging from one to 250 microns, against *Glossina morsitans* and *Glossina swynnertoni*. Spraying should be repeated at 10-day intervals.

The incidence of *Glossina palpalis* can be reduced to a very low level by spraying the vegetation in fly concentration areas by hand with oil solutions or suspensions four times at fortnightly intervals.

Traps have given improved results when their hessian covers have been soaked in kerosene solutions of insecticide. Emulsions have also been used. Screens of hessian which are re-impregnated with insecticide solution every month and are distributed in infested bush at about 10 to the acre, reduce the numbers of *Glossina palpalis* by about 90%.

The treatment of large numbers of cattle in a fly-area with insecticide brought about a slow but considerable reduction in the number of flies; the deposit was renewed once or twice weekly.

Insecticide has been used with promise of success in clearing trains of tsetse flies.

Little work has been done on the use of repellents against tsetse flies. Several repellents have been tested but none appear to be particularly effective.

Clearing. The most effective measure of fly-control is to alter unfavourably the habitat of the tsetse fly by clearing vegetation and removing essential plant associations. Clearing may also be designed to protect man by reducing contact between himself and the fly. Grass is not cut but its growth rather encouraged.

Well-planned clearings give good protection against *Glossina palpalis* and *Glossina tachinoides* which live in dense bush close to water. All fly-belt vegetation is removed from around villages, water holes, etc., to a radius of a quarter to one mile. The protected point is thus put out of the fly's range of vision. Protection may be given at a ford by completely clearing the long, narrow strip of vegetation in which the fly lives, for a mile up and down stream.

Eradication of *Glossina palpalis* and *Glossina tachinoides* can be achieved in an area by partial clearance of the narrow strips of thick, riverine vegetation in which they are found. The thicket and low-branching trees for about ten yards on each side of the stream are felled, only the high-branching trees being spared. This allows the hot wind to penetrate the stream-bed and dissipate the layers of cool, moist air which form over wet sand and pools. An area in which this type of clearance has been carried out must be protected against re-infestation from outside by a mile-deep barrier at the periphery which flies cannot cross. This barrier is achieved by complete clearance of the last (outer) mile of all stream-systems leaving the area.

The most satisfactory form of clearance against *Glossina morsitans* and *Glossina swynnertoni* whose habitat is relatively open, is discriminative clearing directed against a few species of trees upon which the flies are in some way dependent. In one area only about 6% of trees were felled in woodland savannah country, with excellent results in the elimination of fly. The destruction of thicket, for example, by late burning, and the extension of grass is also of value. Some degree of protection may be given to travellers by roadside clearings, carried to a distance of 50–300 yards from the road.

It is important that all clearings should be maintained by annual slashing. Human settlement is a valuable measure of consolidating what has been cleared. Flies cannot live in these areas, and whole populations have been moved into them for safety.

Control points at the exit from an infested area may be established to rid travellers and vehicles of fly by the use of insecticides etc., used, for example, in trains, which may transport flies in very large numbers, particularly on the outside of, and below goods waggons, but also waggons and in passenger compartments.

Game destruction is not effective against *Glossina palpalis* and *Glossina tachinoïdes*, which feed indiscriminately on domestic animals, man, crocodiles, large reptiles etc., in addition to game. It has a place, however, in dealing with the game tsetse *Glossina morsitans* and *Glossina swynnertoni*. Reduction in the number of animals reduces the number of blood meals available so that the fly tends to die out.

Trapping has a place both in the assessment of the effectiveness of control schemes and for local control of flies. Its effectiveness for local control is not absolute. When trapping is used as the only method of control, for example, in sacred sites, it greatly reduces but does not eliminate the fly. An effective trap is a moving screen carried by a fly-boy who catches with a net any fly that alights on it; big catches may be obtained. A framework of light wood, roughly triangular in section, with a flat top, is covered with hessian cloth except for the top and a narrow open slit along the bottom. The top is 6 ft. long by 3 ft. wide and the sides converge below to leave a slit about 3 inches wide. A wire gauze cage is fixed to the upper surface, the gauze so arranged that flies which enter it from below cannot return to the hollow body of the trap. Such traps are best hung on the sunny aspects of the flies' haunts. The effectiveness of such a trap may be increased by impregnating the hessian with insecticide, as described above.

GREY-BOTTLE OR GREY FLESH FLIES (in part) (*Sarcophaga*). These largish flies have a world-wide distribution.

Diseases caused and transmitted by *Sarcophaga* **Grey-bottle flies.** SEMI-SPECIFIC MYIASIS; the larvae occasionally invade wounds, sores, and the vaginal orifice. They are occasionally recovered from the faeces of persons suffering from intestinal disturbance. The adult flies are MECHANICAL VECTORS OF PATHOGENIC ORGANISMS to human food. Plague bacilli have been recovered from the vomit drops of *Sarcophaga carnaria*.

Recognition of *Sarcophaga* **Grey-bottle flies.** They are medium to large, rather bristly (especially the males), powerfully built grey flies, and nearly related and somewhat similar in general appearance to *Calliphora* bluebottle flies. The aristae are plumed on both sides and the wing venation is similar to *Calliphora*. The eyes are a bright red colour in living flies, and the thorax has characteristic dark stripes. The abdomen is diagnostically chequered with black and silver squares and is relatively elongate.

Habits of *Sarcophaga* **Grey-bottle flies.** The adults are greatly attracted to dead animals, excreta and rotting garbage. (The larvae of some species parasitize other insects). They frequently enter houses in search of food.

Life-history of *Sarcophaga* **Grey-bottle flies.** The females of most species deposit *first stage larvae* (often several at once) on the breeding food supply, and not eggs as does *Calliphora*. Otherwise the life-history is very similar to that of bluebottles.

GREY-BOTTLE OR GREY FLESH FLIES (in part) (*Wohlfahrtia*). These flies are not so widely distributed over the world as *Sarcophaga*, being absent in Great Britain and northern Europe.

Diseases caused or transmitted by *Wohlfahrtia* **Grey-bottle flies.** SPECIFIC and SEMI-SPECIFIC MYIASIS; the larvae of some species will invade unbroken skin and cause extensive damage to the underlying tissues. The adults may convey pathogenic organisms to human food.

Recognition of *Wohlfahrtia* **Grey-bottle flies.** These flies, though very similar in general appearance to *Sarcophaga*, may be easily differentiated by the *bare* aristae and by the pattern on the abdomen, which is of rounded black spots on a pale grey background (compare with *Sarcophaga*, above).

Habits of *Wohlfahrtia* **Grey-bottle flies.** These are very similar to *Sarcophaga* in their liking for carcasses, faeces, garbage and human food. The various species exhibit some differences of taste.

Life-history of *Wohlfahrtia* **Grey-bottle flies.** The species concerned in causing myiasis all deposit first stage larvae, not eggs. Some will deposit large numbers of larvae in wounds and sores. Because of their relatively large size the fully-grown maggots of *Sarcophaga* and *Wohlfahrtia* may rapidly cause disfiguring or crippling damage to a living host. The life-history in garbage is similar to that of *Calliphora*.

BOT AND WARBLE FLIES. These are found all over the world where such grass-eating animals as horses, cattle and sheep are present to act as hosts. The female flies deposit their eggs in such a position on the host that the larvae will find access to the living tissue and develop there until the time arrives for pupation, when they make their way to the exterior of the host and drop to the ground, where the puparium is formed and the adult fly finally emerges. The different species of bot-flies each parasitize their chosen kind of host—horses, sheep and so on. All bot and warble flies have characteristically bulky heads and eyes and exhibit great reduction of the mouth parts. With the exception of the horse bot-fly (*Gasterophilus*), the wing venation shows a noticeable relationship host—horses, sheep and so on. All bot and warble flies have characteristically bulky heads and eyes and exhibit great reduction of the mouth parts. With the exception of the horse bot-fly (*Gasterophilus*), the wing venation shows a noticeable relationship to that of *Calliphora* the common bluebottle fly. The ovipositor is usually a large, complicated, hinged mechanism constructed for depositing eggs or larvae on the host animal in the quickest possible time. This is necessary because the herbivorous hosts usually dislike the attentions of these flies and become extremely restive in consequence.

THE HORSE-BOT FLY (*Gasterophilus*). These are found wherever horses, mules, zebras and wild asses occur.

Diseases caused by Horse-bot Flies. SPECIFIC MYIASIS is not uncommon, especially in warm countries, and occurs after a female fly has glued an egg or eggs to human body-hairs, usually on the arms. The larvae when they hatch may penetrate the unbroken skin into the subcutaneous tissue and very slowly develop there for several months, causing a creeping eruption which is seen as a tortuous, inflammatory line. This may give rise to intense itching and secondary infection by scratching. The larva is never able to mature in human beings, however, and as it does not bore its way out it will finally die in the subcutaneous tissue unless previously excised.

Recognition of Horse-bot Flies. The rather large adult flies are clothed in dense fine brown fur and in appearance resemble large hive-bees. There are no bristles. The antennae are situated in a deep facial pit and the aristae are bare. The mouth parts are very poorly developed. The fourth longitudinal vein in the wing proceeds straight to the wing-margin and rather away from the third vein. This is somewhat similar to the venation in *Fannia*.

Habits of Horse-bot Flies. The adult flies appear to be unable

123

to feed. They may be seen in summer-time hovering about the legs of horses grazing in the fields.

Life-history of Horse-bot Flies. The females glue eggs to hairs around the mouth or fetlocks of horses. These are normally licked into the alimentary canal of the animals where the larvae develop by burrowing with their mouth parts into the lining of the intestine. When mature the larvae pass out in the horse's dung and pupate in the ground. The life cycle takes about a year.

THE SHEEP-BOT FLY (*Oestrus ovis*). These are found wherever there are sheep or goats.

Disease caused by the Sheep-bot Fly. SPECIFIC MYIASIS; the adult female flies occasionally deposit a first stage larva (not an egg) in the human eye. This causes immediate pain, and OPHTHALMO-MYIASIS, as the larva burrows into the tissue by means of its very large mouth hooks. Larval invasion of the human nasal passage and pharynx is not uncommon in sheep-country.

Recognition of the Sheep-bot Fly. They are grey-brown flies, a little larger and much stouter than the common house-fly. The head is large and swollen in appearance; the antennae are exceedingly small and the mouth parts are absent. The wings are longer than the abdomen and the body is covered with a fine down. The fourth vein *joins* the third before reaching the wing margin.

Habits of the Sheep-bot Fly. The females hover near the heads of sheep and now and again dart in to deposit a larva enclosed in a drop of milky fluid in the nostril. In human beings the eyes or the nasal openings are usually attacked.

Life-history of the Sheep-bot Fly. Larviposition takes place during the summer and autumn. If unexcised the mature larvae will emerge and fall to the ground in the following spring. Here they pupate, the adult flies emerging some three to seven weeks later.

THE HUMAN-BOT FLY (*Dermatobia hominis*). This fly occurs in Central American and northern Neotropica.

Disease caused by the Human-bot Fly. SPECIFIC MYIASIS; single larvae invade the living tissue and cause boil-like eruptions.

124

The lesions are somewhat similar in appearance to those made by the tumbu fly *Cordylobia anthropophaga*.

Recognition of the Human-bot Fly. The adult flies (about 15 mm. long) are of a brownish grey colour, with the abdomen of a bluish tint and with the legs and face orange-yellow.

Habits and life-history of the Human-bot Fly. The female fly captures a mosquito, biting stable fly, or other biting insect when it is flying, and, holding it firmly against her underside, glues a bunch of eggs to the lower surface of the prisoner's abdomen. The captured insect is then released. When it settles on a human being, one or more maggots of the bot-fly will be stimulated to leave their egg cases and burrow into the skin of the new host. The anterior parts of the larva characteristically enlarge as maturity approaches, while the tail segments remain thin and attenuated and provide for respiration by keeping the posterior spiracles in contact with the atmospheric air.

CATTLE-BOT FLIES OR WARBLE-FLIES (*Hypoderma*). Species of this genus will be found wherever wild or domestic cattle and ruminants are available.

Disease caused by Cattle-bots. Individual larvae not uncommonly invade man and burrow by way of the deep tissues to the neck or head regions. Here they tend to form appearing and reappearing tumours ('warbles') which may be serious or not according to their position. About the eyebrows or eyelids is a not unusual place. If the larvae occur in subcutaneous tissue a creeping eruption results.

Recognition of Cattle-bot Flies. The adults are large hairy flies which are often mistaken for bumble-bees. The antennae are situated in a deep pit and the aristae are bare. The mouth parts are vestigial and the wing venation is not unlike that of *Calliphora*, although the veins are rather crowded towards the front margin.

Habits of Cattle-bot Flies. The females glue their eggs to the body-hairs of ruminants (cud-chewing animals). The larvae, when they hatch out, bore through the hair follicle to reach the living tissue.

Life-history of Cattle-bot Flies. Several hundred eggs are glued to the hairs of the hosts. The larvae, having penetrated the hair follicle, start to bore through the deeper tissues against the pull of gravity, which in a quadruped finally brings the maggot to just below the hide on the back of the animal. The bot-grub seems to be nonplussed by the upright position of man, which results in its wandering from point to point and finally settling in the head area. When mature, the larva will emerge through a hole in the warble—or final tumour in the case of man—and drop to the ground. The grubs are then about one inch in length and have an extremely tough skin. Pupation takes place in the soil. The life cycle requires one full year.

CHAPTER 8

COCKROACHES

Order *DICTYOPTERA*

(by N R H Burgess)

Cockroaches are acknowledged to be one of the most successful of all insect groups. They are among the most ancient of all winged insects and have changed little in appearance during some 250 million years. They are closely related to the mantids which are included in the same Order, and were until recently grouped with the grasshoppers and crickets in the Orthoptera.

There are nearly 4,000 species of cockroach, distributed throughout most parts of the world. The vast majority are of little or no significance to man. Many wild species are diurnal in habit; a large proportion live in tropical forests; some are semi-aquatic, others burrowing or wood-boring; some live in caves and a few live commensally with other insects such as ants. They vary considerably in size, the smallest being only 5 mm. in length and the largest as long as 9 cm. Typically they are oval flattened creatures, with a clearly defined head, large pronotal area on the thorax, three pairs of legs and a pair of long antennae. Most species have two pairs of wings. Cockroaches are exopterygote. Some time after mating the female lays a purse-like egg-case or oötheca containing from 12 to 48 eggs according to species. From these hatch nymphal insects looking very much like miniatures of the adult but without wings. By passing through as many as twelve instars the insect becomes an adult.

About 50 species of cockroach have acquired the habit of domestication to a greater or lesser extent, and some have followed man to every part of the globe, a few becoming widespread and important domestic pests.

Diseases caused or transmitted by pest Cockroaches. Evidence is accumulating that cockroaches may act as mechanical vectors and efficient reservoirs of disease, especially of the excremental type. Their feeding habits are somewhat similar to those of flies though they are omnivorous in their diet. In tropical and subtropical areas they are perhaps not as mobile as flies, but in temperate regions they are prevalent for the whole year, whereas flies are only common in the warmer season. Cockroaches will feed readily on sources of infection such as human faeces, refuse, surgical swabs, sputum etc. and are often found frequenting sewers, drains and dustbins. They will also habitually infest kitchens, dining rooms, and food stores in hospitals, hotels and domestic dwellings where they walk over and defaecate on food intended for human consumption, or pass on pathogenic organisms in vomit drops or from their legs and body surfaces.

Cockroaches in their natural domestic environment have been found to be contaminated with at least 40 different species of bacteria pathogenic to man, as well as helminth eggs, fungi and poliomyelitis virus.

Recognition and life-history of pest Cockroaches. Three species of pest cockroach found in temperate regions are *Blattella germanica* the German cockroach, *Blatta orientalis* the oriental cockroach and *Periplaneta americana* the American cockroach.

The German cockroach or steamfly (Fig. 45) is about 12–15 mm. long and yellowish brown in colour. It moves quickly and is an adept climber. Both sexes have well-developed wings. This species occurs in warm and humid environments. The egg capsule is carried by the female until just before hatching and produces up to 30 young nymphs, each female laying 4 or 5 oöthecæ during her life-time of about 9 months. Adults are fully grown in 3–4 months.

The oriental cockroach (Fig. 46), sometimes erroneously called the 'black beetle', measures 25–30 mm. in length and is a very dark brown in colour. The wings of the female are only just apparent, while those of the male are larger but still shorter than the body. It tends to be rather slower than the German cockroach and not such a good climber. It seems to prefer cooler drier situations though it may often be found infesting the same areas as the German cockroach which seems to be replacing it in many areas.

The female oriental cockroach will deposit the oötheca a few days after it is formed rather than carry it around, and will lay up to five during her adult life of about 5 months. As many as 16 young

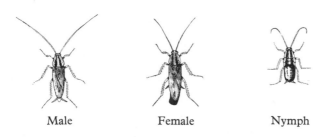

Male   Female   Nymph

Figure 45   *Blattella germanica*

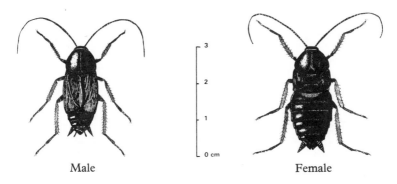

Male   Female

Figure 46   *Blatta orientalis*

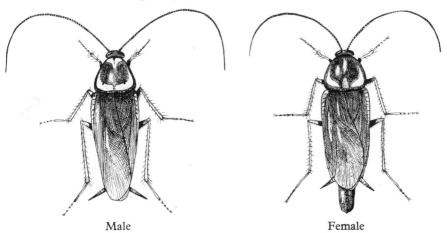

Male   Female

Figure 47   *Periplaneta americana*

(Figs 45–47 drawn by S. N. McDermott)

nymphs will hatch from each capsule after 6–12 weeks and will become adult in about 10 months.

The American cockroach (Fig. 47) measures up to 40 mm. in length and has very long antennae. It is light brown in colour with yellowish markings on the pronotal part of the thorax. It is less common than the other two species, though found in warm humid sewers, warehouses and on board ships. The female will deposit her oötheca at an early stage, from which hatch up to 14 nymphs. As many as 30 capsules may be laid by a female during her adult life of some 15 months. Nymphs will take as long as 18 months to reach maturity.

Habits of Cockroaches. Domestic pest species of cockroach are essentially nocturnal in habit, and although present in large numbers, may not be seen in offices, canteens etc. used only during the day. Inadvertent wanderers, cast skins or dead bodies of cockroaches, structural repairs to the building or room, or the tell-tale brown faecal streaks are a sure sign of infestation, and a visit to the premises after dark will certainly provide confirmation. During the day they will hide in cracks and crevices, behind panelling and furniture, in ducts, and under shelving etc.

Control of Cockroaches. The complete, long-term eradication of an established infestation of pest cockroaches is often difficult and expensive in terms of man-hours and insecticides, but it should be possible without excessive expenditure to keep any infestation at an acceptable level so that the presence of cockroaches is not apparent or detectable during normal activities in the premises. Good and necessary standards of cleanliness, hygiene, maintenance and inspection should ensure that any premises are kept free of cockroaches, and that any minor infestation which is introduced is dealt with immediately and effectively.

The *design* of the premises, siting of equipment and furniture etc., is important in the control of cockroaches. For example, equipment in kitchens should be placed at least a foot away from the wall so that spaces behind can be cleaned and do not provide harbourages for cockroaches. Hollow wall-cavities and false ceilings should be avoided if possible and access should be made available to heating and ventilation ducts, so that these spaces can be treated with insecticides.

Conditions of *hygiene* are vital in the maintenance of cockroach-free premises. Food should not be left out at night, scraps should be cleared away, surfaces cleaned and dustbins emptied, so as not to encourage cockroaches. It is pointless to attempt cockroach control in any premises where conditions of hygiene and cleanliness are poor.

Once the above faults have been rectified and standards maintained, cockroach control can best be achieved by the use of *insecticides*. Cracks and crevices, floor-wall-ceiling margins, door and window frames etc. should be treated with a residual insecticidal spray or lacquer. Dead spaces such as ducts and hollow ceilings should be treated with insecticide (boric acid powder which acts as a stomach poison is useful in these areas), and in addition the whole site may be fogged with a knockdown insecticide which will excite the insects and encourage them to run over residually treated surfaces. Where possible a complete block should be treated, above, below and on either side of the infested area, so that cockroaches do not leave one room and infest another. It is important that floors and walls are not washed down for several days after treatment since this will remove the insecticide. Once the area has been treated, regular monitoring at night should be undertaken, and further treatment carried out when necessary or advised by the insecticide manufacturer. Maker's instructions should always be read before using the insecticide, and recommendations, particularly safety precautions, adhered to.

CHAPTER 9

INSECT VECTORS IN THE EUROPEAN REGION INCLUDING THE BRITISH ISLES

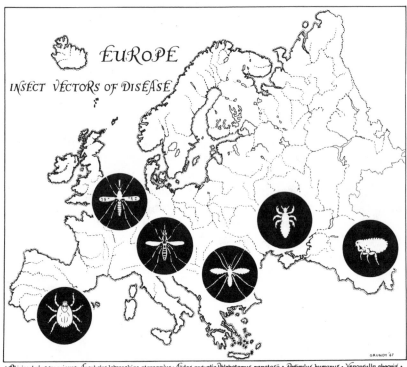

* *Rhipicephalus sanguineus* · *Anopheles labranchiae atroparvus* · *Aedes aegypti* · *Phlebotomus papatasii* · *Pediculus humanus* · *Xenopsylla cheopis* ·

The study of present-day vectors of disease in Europe including the British Isles, gives rise to certain difficulties.

There are a few people still living who have spoken to witnesses of malaria, typhus, relapsing fever, and cholera epidemics in the British Isles (*circa* 1850); while more recollect European epidemics of insect-transmitted malaria, typhus, relapsing fever, typhoid, plague, cholera, and so on in the days before the use of modern insecticides.

The present danger to the populations concerned lies in the possible reintroduction of controlled diseases to territories still suitable for their arthropod vectors, during periods of disorganisation brought about by unstable political conditions. Also, the non-immune individual is always at risk when using modern forms of transport such as the aeroplane and motor car to reach out of the way places where insect-borne diseases still maintain an endemic foothold.

This danger of the individual contracting diseases when travelling in Europe, particularly in view of the recent large-scale movements of tourists and others to and from Africa, is greater than is popularly supposed.

A basic danger is that when a control or eradication programme against disease vectors reaches its target, it is usually discontinued, which allows the vector species gradually to return; and so, as for instance in European malaria, to provide conditions for a local outbreak, if only one suitable human carrier becomes available to these vectors. In Sardinia, for example, malaria still persists in spite of the eradication project.

There are several other conditions that hide the potential danger, such as (i) the public preferring not to be reminded of any danger, (ii) commercial interests avoiding any mention of diseases contracted in holiday or trading areas, (iii) unsuspecting or hard-worked clinicians not always recognising imported diseases for what they are, and even if they do so, of not always reporting their findings.

The main communicable insect-borne diseases of the European region are as follows:

1. Classical louse-borne exanthematous or epidemic typhus. This is caused by *Rickettsia prowazeki* the type species of the genus *Rickettsia*, and transmitted from an infected human to another human in the faeces of an infective human louse *Pediculus humanus*, which is specific to man. The rickettsiae in these faeces may be scratched into the skin by the human host or may become dried and air-borne, and so able to enter human respiratory passages.

Vector species.—*Pediculus humanus* Linnaeus, 1758 (*pediculus*, attendant foot slave) containing *Pediculus humanus* var. *capitis* Degeer, 1778 the head louse, and *Pediculus humanus* var. *corporis* Degeer, 1778 the body louse.

Vector synonyms.—*Pediculus vestimenti* the clothing louse.

Vector and disease notes.—Until fairly recently there were frequent epidemics of louse-borne typhus in Ireland, Spain, the Balkans and Eastern Europe generally. Until 1870 the death rate in England and Wales alone was approximately 4,000 per year. Head lice as well as body lice are important typhus vectors as they foul the head hairs with their excrement which if infective may be scratched into the skin of adjacent persons and also become air-borne. *Rickettsia prowazeki* enters the cytoplasm of the endothelial cells lining the gut of the infected louse and there multiplies until the cells are destroyed, when the fragments and the organisms join the infective faeces. At the commencement of a blood meal, the first of the host's blood is typically passed through the hind gut of the louse and voided through the anus. Later dry faeces will be passed, often joined together in strings. It is the faeces of infected lice that form the most important route of epidemic typhus to man, as the saliva in the bites is much less important. It is curious that lice tend to leave a febrile patient and seek human hosts with a more normal temperature, thus spreading (if infective) the rickettsiae to neighbouring persons.

The bites of lice do not affect some people, but the majority react with irritating papules which may become secondarily infected and cause itching, urticaria and serious loss of sleep. Brushes, combs and head-gear may spread head lice, and under-clothing or outer body clothing spread body lice.

It should also be noted that cases of tick-borne typhus or fièvre boutoneuse occur each year in southern Europe.

2. Epidemic louse-borne relapsing fever. This is caused by *Borrelia recurrentis*, a pathogenic spirochaete which is transmitted from an infected human to another human, in the blood-fluid but not in the saliva or faeces of an infective human-louse *Pediculus humanus*. Man typically acquires the disease, when the integument of an infected louse is ruptured or torn between the finger-nails or teeth, which enables the spirochaetes in the louse-blood to penetrate the adjacent skin tissue, mucous membrane or conjunctiva.

Vector species.—*Pediculus humanus* Linnaeus, 1758 containing the head louse and the body louse.

Vector and disease notes.—Louse-borne relapsing fever epidemics no longer occur outside the Middle East and north-east Africa. They were recently common, however, from Ireland to eastern Europe. They are caused by the spirochaete *Borrelia recurrentis*, sometimes named *Borrelia obermeireri*, which after being ingested from the peripheral blood of an infected person leave the digestive tract for the haemocoele so that neither the louse saliva nor the faeces are ever infective. Six days or so after the infecting meal the blood of the louse contains numerous spirochaetes which if passed on to thin human skin through a wound in the louse integument, may transmit the pathogens to man.

Tick-borne relapsing fever cases are reported annually in Spain.

3. Mosquito-borne benign tertian malaria. (It is now claimed by the WHO that malaria, for practical purposes, has been eradicated from continental Europe.)

This is caused by the protozoan parasite *Plasmodium vivax*, the main species responsible for malaria in the temperate region. The parasite is transmitted from an infected human to another human in the saliva of a female *Anopheles* mosquito while she imbibes a blood meal. The vector mosquito acquires the pathogen by piercing the skin and sucking the peripheral blood of a person harbouring gametocytes or male and female forms of the parasite, which proceed to mate and multiply in the mosquito and finally colonize her salivary glands. There are some nine or more possible vectors of malaria in Europe, mostly members of the dominant temperate region *maculipennis* complex.

Vector species.—*Anopheles maculipennis* Meigen, 1818 = *Anopheles maculipennis* var. *typicus*. *Anopheles labranchiae atroparvus* Van Thiel, 1927 = *Anopheles maculipennis* var. *atroparvus*. *Anopheles maculipennis messeae* Falleroni, 1926 = *Anopheles maculipennis* var. *messeae*. *Anopheles labranchiae* Falleroni, 1926 = *Anopheles maculipennis* var. *labranchiae*. *Anopheles melanoon* (not considered a vector). *Anopheles melanoon subalpinus* (Hackett and

135

Lewis, 1935 (considered a vector)). *Anopheles saccharovi* Favre, 1903 = *Anopheles maculipennis* var. *elutus*. *Anopheles claviger* (Meigen, 1804). *Anopheles plumbeus* Stephens, 1828. *Anopheles superpictus* Grassi, 1899. *Anopheles algeriensis* Theobald, 1903.

Vector and disease notes.—For purposes of describing the distribution of the *Anopheles* vectors of malaria, Europe may be likened in shape to an approximate oblong (with Iceland in the north-west corner, where, however, there are no *Anopheles* or malaria).

The north is predominantly cold due to Arctic influences.

The south is predominantly warm, due to its more southerly latitude and African influences.

The east is predominantly dry due to being land-locked far from any oceans.

The west is predominantly cool and rainy, due to the north to south Gulf stream that washes its shores and the alternating 'lows' and 'highs' of the prevailing westerly winds blowing onshore from the North Atlantic.

This makes for two climatic zones (i) a northerly cold-temperate zone north of the Alps, and (ii) a southerly warm-temperate zone south of the Alps to the Mediterranean sea.

Plasmodium vivax and *Plasmodium falciparum* compete in the warm-temperate zone, south of the Alps, but *Plasmodium vivax* is dominant in the northern temperate zone.

The difficulty for malaria parasites as the latitude becomes more northerly is the cold winter period joining with unfavourable spring and autumn months, when they must hibernate either in the body of a suitable mosquito or in a human reservoir. Malaria parasites typically cannot develop within a mosquito if the temperature is below 60°F though parasites living commensally in northern species such as *Anopheles labranchiae atroparvus* are more liable to maintain their effectiveness through a cold winter than the more southerly forms. *Atroparvus*, however, only partially hibernates, that is, it feeds indoors at irregular intervals on domestic animals (chiefly cattle and pigs) or 'domestic man', for it always chooses the corners of warm dark stables, byres or cottages in which to pass the winter period. *Anopheles maculipennis messeae*, on the other hand, spends the winter in any cold but reasonable shelter and

136

hibernates completely, living on its stored fat-body without augmenting this with blood meals. This difference in the habits of two almost identical mosquitoes explains why *Anopheles labranchiae atroparvus* is able to cause 'house-malaria' during the winter, especially when in proximity to children who are good gametocyte carriers, while *Anopheles maculipennis messeae*, though closely related, never does so.

The main distribution of *atroparvus* follows brackish waters inland of the coast line. A physical relief map will show these typical areas lying in a diagonal strip from Spain to Finland. Similar conditions such as occur in the Danube delta on the Black Sea also form breeding areas associated with *vivax* malaria.

Messeae is more centralized and widespread in Europe than *atroparvus* because it is mainly a fresh water breeder, which allows it to penetrate far inland among river basins and marshes. It prefers the blood of cattle to that of man during the warm months and hibernates completely in the cold months, and so is only a vector of malaria where dense human populations occur near the breeding marshes in the absence of cattle, as in certain parts of Hungary. It does not breed to the south-west beyond the Balkans.

Anopheles labranchiae labranchiae links the northern varieties of *maculipennis*, that is, *atroparvus* and *messeae*, with the warmer western lands of Europe such as Italy and Spain where it is still a dangerous potential vector. This species overlaps with the eastern variety of *maculipennis* now named *Anopheles saccharovi* which is the chief vector south and east of the Alps where it ranges as far as the Caspian. It breeds mainly in brackish water and its eggs are remarkable in typically lacking the side floats that are so characteristic of *Anopheles* mosquitoes. *Anopheles saccharovi* is a dangerous vector of malaria over most of its range.

Anopheles melanoon subalpinus is found from Switzerland to the Caspian and is a vector of a severe malaria. It partially hibernates like *Anopheles labranchiae atroparvus* and is thus also able to cause 'winter house-malaria'.

Anopheles claviger (which is not a *maculipennis* variety) is very widely distributed in both the northern and southern countries of Europe. It causes local outbreaks of malaria especially where cisterns of water suitable for breeding are placed below houses.

Anopheles plumbeus is widely distributed in northern Europe,

137

where it is primarily a tree-hole breeder and has in the past been thought responsible for isolated cases of malaria in large towns such as London.

Anopheles superpictus is an east of Italy foot-hill mosquito, breeding in cool moving water often free of vegetation but with plenty of sunshine, and is responsible for unstable areas of endemic malaria mainly in summer, which instability is probably due to a natural preference for horses and cattle when these are available.

4. Dengue, dandy (*denguero*—one with affected gait) **or break-bone fever** (term commonly used *circa* 1870). This is caused by a group B arbovirus with six known strains, which is transmitted from an infected human to another human in the infective saliva of a female *Aëdes aegypti* mosquito during a blood meal.

The vector requires some eight to fourteen days to incubate the virus to the infective stage, after which its saliva remains infected for the duration of its life, which may be several months.

Dengue has a wide distribution in warm countries, and follows the range of the *Aëdes aegypti* mosquito. Cases occur every year in Spain, Italy, Austria (Vienna had a large epidemic in 1941) and Yugoslavia; in fact, any country bordering the Mediterranean may suffer sudden outbreaks of this disease.

Vector species.—*Aëdes aegypti* (Linnaeus, 1762).

Vector synonyms.—*Stegomyia fasciata.* Tiger mosquito, Yellow fever mosquito.

Vector and disease notes.—Since *Aëdes aegypti* is adversely affected by cold conditions, dengue epidemics invariably occur in the hottest part of the year when the mosquito is most numerous. Extensions of the illness beyond the normal geographical range take place during a specially hot summer and autumn. Most epidemics and the sporadic cases that originate them occur in coastal areas but may also appear inland if conditions there suit multiplication of the mosquito. The black and white *Aëdes aegypti* females are semi-domestic and very small, although larger than the males, and they fly more silently than, for instance, the common brown house-mosquito *Culex pipiens* (*pipio*, the piper). They bite in the shade at any time of the day, or by night, and, for preference, quietly attack the back of the neck or ankles.

The eggs are invariably laid on or just above the water line of any tin, pot or other suitable container. Inside these eggs, in a humid atmosphere, first stage larvae will develop within 48 hours, but these will not emerge until rising water in the receptacle where they were laid stimulates them to break open the egg-shell and swim free. During drought conditions, however, the egg may remain viable for months, even years, without hatching. There are four aquatic larval stages, as with all mosquitoes, with a moult between each, and a comma-shaped active pupal stage in which are developed the wings, legs, proboscis etc. of the adult.

5. Sandfly fever, Phlebotomus fever, papataci fever, or three-day fever. This is caused by a Group C arbovirus known as the sandfly-fever virus, and is transmitted from an infected human to another human only in the saliva of an infected *Phlebotomus papatasii* sandfly during a blood meal.

The vector requires seven to ten days from the initial meal on a first day fever patient, to incubate the virus to the infective stage. Occasionally the organisms are passed to the next generation of sandflies, usually by the larvae eating the dead bodies of infected female sandflies.

Vector species.—*Phlebotomus papatasii* (Scopoli, 1786).

Vector synonym.—Golden coloured dry-area sandfly.

Vector and disease notes.—Sandflies are very small in size, and although ranging as far north as Paris, are, like the *Aëdes aegypti* mosquito, adversely affected by cold, and therefore must have a warm to hot climate in order to multiply. However, even the 'dry-area' sandflies require a high humidity for most of the day, which, in dry surroundings, they endeavour to find in microclimates such as drought cracks in heavy soil, old pointing in buildings or animal burrows.

Sandflies typically feed at dusk and dawn when the stillness of the atmosphere interferes least with their weak powers of flight. The interior of a room suits their biting habits, for apart from the disturbing action of an electric fan the air is usually quiet enough for them to control their light bodies in flight. *Phlebotomus papatasii* is markedly anthropophilic, and causes a sharp pricking sensation when puncturing the skin to initiate a blood-meal.

139

Usually sandflies bite only in ground-floor rooms, as they dislike flying high, but if drain-traps and pipes connect various floor levels, then sandflies may be found at the top of high buildings.

The golden coloured sandfly is so well camouflaged when looked for against a neutral background that often only the intensely dark eyes can be seen. The covering of all parts with fine hairs, including the wing veins, causes sandflies to be silent in flight, in contrast to most mosquitoes.

Ordinary mosquito netting will not prevent the entry of unfed female sandflies, though when gorged with blood they find it difficult to push their way out. The use of sandfly netting (45/46 mesh), however, is not advised as it is almost impossible to sleep under it, due to lack of air movement in the confined space.

The life-cycle takes place below ground in deep crevices or piles of rubble. The four stages of the tiny maggot-like larvae are, therefore, without eyes. The period required from egg to adult is about eight weeks, and the adults may live up to three weeks.

Man appears to be the only reservoir of the sandfly fever virus (unless the passage of the virus to the next generation enables the sandfly itself to be a reservoir), and usually suffers two short seasonal epidemics in endemic areas, the first coinciding with the hot weather of summer and the second in autumn, the eight weeks or so between these periods being the time required to produce a new generation of sandflies.

6. Cutaneous leishmaniasis or oriental sore. This is caused by a protozoon known as *Leishmania tropica*, and is transmitted from the indurated edge around a leishmanial sore or ulcer on the skin of an infected animal (*e.g.* dog) or human, to another human by the biting mouth parts of a female *Phlebotomus papatasii*, during a blood meal, the incubation stage in the sandfly taking approximately 8–21 days. This species of sandfly is house-frequenting and therefore is in constant contact with man. The dry-area female *Phlebotomus papatasii*, the distribution of which is centered on north Africa, though prevalent in southern Europe, is almost indistinguishable from the other dry-area vector *Phlebotomus sergenti*, which is centered on the Near and Middle East, and India. Indeed, although the disease is frequent in southern Europe, and occasional epidemics occur, the main focus is just outside the boundaries of Europe in Turkmenistan where there is a permanent reservoir among the wild rodents (*e.g.* the gerbil *Rhombomys opinus*) often in uninhabited areas. The endemic countries of Europe include practically the whole of the Mediterranean littoral and islands, from Spain to the Black Sea.

Vector species.—*Phlebotomus papatasii* (Scopoli, 1876). *Phlebotomus sergenti* (Parrot, 1917).

Vector and disease notes.—The above two species are markedly anthropophilic but will feed on other animals when human blood is not available. Sandflies, in nature, do not usually live longer than a fortnight, and therefore need to become infective and to pass on the infection as soon after maturity as possible. The first act of oviposition, during which the eggs are expelled with considerable force, often kills the parent, and any condition of dryness is inimical to successful breeding at any time from egg to adult.

 Phlebotomus perfiliewi (Parrot, 1930) has been reported as a vector of Oriental Sore in Italy, where it breeds in manure piles. If dwellings are situated some distance from these breeding sites, however, then there appears to be no transmission.

7. Visceral leishmaniasis or kala-azar. This is caused by a protozoan known as *Leishmania donovani*, and is transmitted from the peripheral blood of an infected human (and possibly infected dogs in the Mediterranean region), or, from post-kala-azar dermal lesions, to another human, by the biting mouth parts of a female *Phlebotomus perniciosus* sandfly. The incubation stage in the sandfly takes approximately 7 days.

This disease does not commonly overlap the territories of the dry-area vector sandflies but in some places there is a superimposition. In Europe sporadic cases may occur along the southern coast of Spain, the south of France, lower Italy, Sicily, Yugoslavia and Malta, Crete, Hydra and other Greek islands, together with various seaports. In these areas young children and infants acquire the disease much more frequently than adults.

The sandfly vector *Phlebotomus perniciosus* has not tufts on the dorsum of the abdomen as has *Phlebotomus papatasii*. It also differs from this sandfly by not readily entering houses.

Vector species.—*Phlebotomus perniciosus* Newstead, 1911.

Vector and disease notes.—Kala-azar in Europe is almost confined to the very young and is therefore known as the infantile type. The infective organisms are not conveyed to man in the saliva of the sandfly, as are the parasites of malaria in the mosquito, but infected directly from the pharynx and food channel of the proboscis, following pressure originating in a blockage of the oesophagus with leishmanial leptomonads. In the human skin, these immediately lose their flagella and become rounded in shape, and in this form are ingested by the host's tissue macrophages, where they proceed to multiply and eventually burst the cell. The new crop of Leishman-Donovan bodies will then be taken up by other macrophages.

8. Bubonic plague, pneumonic plague, plague, Black Death. This disease is caused by a bacillus *Yersinia pestis* which is typically transmitted from semi-domestic and domestic rats, *Rattus norvegicus* the brown ratand *Rattus rattus* the black rat, respectively, which have died of rodent-plague, to a human, by the infected mouth parts of a plague-rat flea, *e.g. Xenopsylla cheopis.*

Vector species in Europe.—*Xenopsylla cheopis* (Rothschild, 1903) the 'tropical plague flea'; *Pulex irritans* (Linnaeus, 1758) the 'temperate region human flea'; and *Nosopsyllus fasciatus* (Bosc, 1800) the European or temperate region rat flea.

Vector and disease notes.—Plague is primarily a zoonosis of wild rodents, and the original plague centre was probably among the tarabagan *Marmota bobac* of Central and East Asia. Chains of the infection through various wild rodent species, at different periods, moved westwards over the mountains of Kurdistan to the Mediterranean; and also eastwards to China, from whence, *via* the trade routes at the turn of the century, they eventually reached the Orient in general, Australia, Africa and the Americas.

The last world-wide pandemic of plague thus started in China about 1894 where it appeared in that year in Canton and Hong Kong. This pandemic is only now reaching its nadir, and it is hoped it will be kept there by the Health Authorities with the aid of insecticides able to destroy the flea populations of plague-carrying rodents. The difficulty of doing this, however, is shown by the fact that plague organisms are still advancing across the countryside of North America, at present in thinly inhabited country, in the living bodies of the ground marmots, squirrels and rats of that vast territory.

In Europe, including the British Isles, plague only appears as brief outbreaks in sea-ports, where infected rats by way of their infective fleas pass the *Yersinia pestis* organisms to a limited number of human beings. Glasgow, Liverpool, Cardiff, Bristol, Hull and London have all suffered such minor attacks in the past. In the Mediterranean there is always a danger of plague spreading from sporadic areas in North Africa which stretch from Morocco to Egypt, as the epidemics of Malta during 1945 and 1946 (22 deaths), Corsica in May to July 1945 (10 deaths) and the instep of Italy in September to November 1945–46 (Taranto—15 deaths), clearly show.

A most curious smouldering of rural plague occurred in East Anglia between 1906 and 1918, when plague was eventually found to be wide-spread among the rats, rabbits, hares and even ferrets over a limited territory. Seventeen villagers died of plague during these twelve years, the deaths being at first attributed to pneumonia. The infection was eventually traced to grain lighters importing plague infested fleas into the area around the River Orwell.

There is still some mystery about every aspect of the transmission of plague. For instance, typical bubonic plague manifests itself by the appearance of classical inguinal buboes or swellings, but at times plague develops a pneumonic form which is not necessarily transmitted by plage-fleas as it becomes droplet-borne and so is passed from person to person in the breath, particularly as there is violent coughing as the disease progresses.

Again, it is known that the most dangerous plague flea is the oriental or tropical rat-flea *Xenopsylla cheopis*, since this flea has a gizzard-valve that blocks more readily with colonizing *Yersinia pestis* than any other species of flea, and therefore struggles to unblock this plug of plague organisms into a bite puncture more frequently than other 'slower to block' rat-fleas; yet plague epidemics may take place in the absence of *Xenopsylla cheopis*, as they are typically fleas requiring heat and humidity for their well-being (they flourish on rats in central-heating systems in Britain if rats are permitted to exist there). *Pulex irritans*, on the other hand, although it is able to multiply under much cooler conditions, is not considered a good transmitter of *Yersinia pestis*, although in some

143

parts of the world, where it is permitted to swarm in rural dwellings, it may, after a large number of the fleas have fed *en masse* on a case of plague, transmit sufficient plague organisms to members of the household and near neighbours as to cause a local epidemic.

Nosopsyllus fasciatus, the temperate region rat flea, similarly, under suitable circumstances, is believed capable of maintaining endemic plague among rodents, although the aid of *Xenopsylla cheopis* is considered necessary for any serious outbreaks of plague to be initiated and maintained.

9. Diseases from human excrement—typhoid, cholera, dysentery, poliomyelitis, infant diarrhoea, helminthiasis. The following list of diseases, caused by various bacterial, viral, protozoal and helminthic pathogens, may, under suitable conditions, be acquired from infected sources, such as human faeces and vomitus, and be transmitted to man by certain semi-domestic or domestic calypterate flies, especially *Musca domestica*, the common temperate region house-fly that frequents domestic premises and in particular kitchens, eating places and lavatories:

Cholera (*Vibrio cholerae*), bacillary dysentery (*Shigella dysenteriae*), amoebic dysentery (*Entamoeba histolytica*), typhoid (*Salmonella typhi*), tuberculosis (*Mycobacterium tuberculosis*), anthrax (*Bacillus anthracis*), helminthiasis (*Taenia solium* ova) etc.

Vector species.—
Musca domestica (Linnaeus, 1761) temperate region house-fly.
Calliphora erythrocephala (Meigen, 1826) temperate region blue-bottle fly.
Calliphora vomitoria (Linnaeus, 1758) temperate region blue-bottle fly.
Fannia canicularis (Linnaeus, 1761) lesser house-fly.
Fannia scalaris (Fabricius, 1764) latrine fly.
Phaenicia sericata (Meigen, 1826) = *Lucilia sericata* green-bottle fly.
Muscina stabulans (Fallen, 1817) non-biting stable-fly.

Vector and disease notes.—The potential danger of the above flies lies in the availability or otherwise of infected material on which they may alight and feed. In Europe generally, the standard of hygiene is high enough to deny regular access to human faecal matter containing, for instance, typhoid bacilli, or carcasses infected with pathogenic organisms such as anthrax.

Semi-domestic and domestic flies, however, are always ready to avail themselves of any relaxation of sanitary standards and to welcome back epidemics of cholera, typhoid, anthrax etc.

There are three main fly-routes for the transmission of disease organisms to man. Firstly, dejecta from their numerous body hairs, which they are continually using as cleaning brushes to get rid of dust or particles of foreign excreta from their typically bristly exterior. This route also includes the paired foot-pads, which being 'adhesive', are well constructed for hanging on to a smooth overhead surface or for collecting microscopic organisms from human or other faeces which are then carried from place to place, including human food.

Secondly, from the vomit-drop or bubble of fluid which is constantly protruded and retracted so as to moisten the delicate lobe-like mouth parts situated at the tip of the folding proboscis, and which is full of whatever organisms the fly has been able to imbibe in previous drinking sessions on any acceptable liquid, including the liquid in faeces.

Thirdly, from the development of pathogens within the long alimentary tract of the fly, so that its faecal deposits may later transmit a multiplicity of these organisms to the person or food of man.

An increase of the fly population as the summer advances tends to run parallel to the numerical increase of their commensal organisms. For instance, flies examined early on may have comparatively few bacteria on their bodies, while later in the season they may be carrying more than six million per fly. It will usually be found that flies contain more than ten times the bacteria within the alimentary tract than on their outside surfaces. Furthermore one pair of house-flies on a refuse tip where there may be few natural enemies together with an unlimited food supply, may, during a warm humid period of twenty-one weeks, produce the fantastic number of 408,582,844,858,520 adult male and female flies.

While it is difficult to incriminate flies in the direct transmission of diseases, because the routes are invariably mechanical rather than by inoculation (as occurs when a female mosquito transmits malaria) yet a diminution of the local fly population by special control measures during epidemics of the excremental diseases invariably coincides with a fall in the number of new cases.

The following few notes may give food for thought: typhoid bacilli acquired by house-flies from infected human faeces are known to survive within the alimentary tract of the flies for more than three weeks after acquisition, and thus, by way of their drowned bodies in milk or water, or other methods of food contamination, may be able to transmit the infection back to man *via* the oral route. Bacillary dysentery, now rare in Britain, was once known as asylum or goal dysentery. In urban areas epidemics have always been affected markedly by the fall of the house-fly population following control operations and the removal of infected faecal material. Bacillary dysentery is a disease of hot summers and autumns in temperate Europe, but in countries such as Iraq which have an extremely hot summer season, the disease ceases when the heat temporarily kills off the flies. Although communicable to all ages, bacillary dysentery, wherever it occurs, is a serious cause of infant and child mortality. By contrast, amoebic dysentery, a protozoal disease prominent in Italy and Greece, which is also transmitted by flies, is rare in children. Man is the sole reservoir of the infection, which is conveyed from one person to another by the contamination of personal contact, ingested food or flies, and the transmission is always by this oral route. Cholera vibrios have been found in house-flies naturally infected during human cholera epidemics, and after an incubation period of some five days within the alimentary tract of the house-flies, were infective during a further week when defaecated on vegetables and fruit intended for human consumption. The infection of man, who is the sole reservoir of cholera, always takes place by the oral route, and house-flies, with their avidness in hot weather for liquid faeces and vomitus, are the most important active vectors of the disease, contaminated water being the chief passive agent. The last pandemic of cholera to reach England was in 1865, and 78 cases of cholera were reported from Ukraine during the Second World War, but the last near-Europe epidemic was in Egypt in 1947, which killed more than 20,000 people before being brought under control.

The poliomyelitis virus has been repeatedly isolated not only from house-flies but also from other filth-feeding flies such as *Phaenicia sericata* the green-bottle fly, for the disease is primarily an infection of the human alimentary tract with only secondary localization in nerve tissue. These isolated viruses remain viable for varying periods, ranging from forty-eight hours to more than twelve days. Tubercle bacilli have been retained within house-flies for four or five days after having had access to infected human sputum.

[In the same context, it is worth remarking on the part played by cockroaches in the transmission of excremental and other disease organisms. While in most areas of Europe flies are seasonal, appearing as adults only during the warmer part of the year, cockroaches, particularly the German cockroach *Blattella germanica* and the oriental cockroach *Blatta orientalis* are present infesting warm domestic situations all the year round. Cockroaches have somewhat similar although omnivorous feeding habits to the fly and have been shown to survive exclusively on human faeces for at least ten weeks. They will take up almost any organism in their environment and in many cases the organism will survive and be passed out in the cockroach faeces on to human food, cooking utensils and other materials from which man may become infected. Potentially pathogenic organisms found in the gut of cockroaches naturally infesting sewers and domestic premises in temperate Europe include *Escherichia coli*, *Klebsiella pneumoniae*, *Serratia marcescens*, *Pseudomonas aeruginosa*, *Shigella dysenteriae*, *Salmonella typhi*, *S. typhimurium* and *S. bovis-morbificans*, as well as four strains of poliomyelitis virus, the fungus *Aspergillus fumigatus*, the protozoan *Entamoeba histolytica* and a number of helminths affecting man—EDITOR'S NOTE.]

CHAPTER 10

CONVERSATIONS

The form of these Conversations has been used for many years by medical officers studying tropical medicine, hygiene and allied subjects at the Royal Army Medical College. The questions and answers are arranged to stimulate interest and test and supplement the knowledge acquired while reading the foregoing chapters.

1. *Write down briefly the characteristics common to all Arthropods* (e.g. crustaceans, scorpions, spiders, ticks, mites, insects).
(1) Bilaterally symmetrical.
(2) Metamerically segmented.
(3) Chitinous, sometimes sclerotized exo-skeleton.
(4) Paired jointed legs, *at least one pair* of which is modified as 'jaws'.
(5) Haemocoele, enclosed by exo-skeleton.
(6) Dorsal tubular heart.
(7) Intermediate alimentary tract with invaginated chitinous fore- and hind-gut.
(8) Central nervous system of two longitudinal ventral nerve trunks.
(9) Respiration by gills (*e.g.* some aquatic forms); lung books (*e.g.* some terrestrial forms); through cuticle (*e.g.* some small parasitic forms); spiracles (*e.g.* all aerial forms).
(10) Sexes separate.

2. *Write down briefly the morphological features which differentiate Arachnids (*e.g. *scorpions, spiders, ticks, mites) from Insects (*e.g. *bugs, lice, fleas, flies).*
Arachnids have their body segmentation divided into a fore-body ('cephalothorax') and hind-body ('abdomen'). (Ticks and mites have these sections fused into one sac-like body.)

The fore-body bears one pair of jaws (chelicerae) and one pair of palps.

Simple eyes may be present or absent. There are four pairs of walking legs.

The hind-body has no limbs and the opening to the gonads is always forward on the ventral surface of the hind-body.

Neither wings nor antennae are ever present.

Insects have their body segmentation divided into head, thorax and abdomen.

The head always has a pair of antennae and three pairs of mouth parts, typically a pair of compound eyes, and sometimes simple eyes (ocelli).

The thorax is of three segments and always has three pairs of legs and usually two pairs of wings.

The abdomen may have up to eleven segments but never bears legs.

The opening to the gonads is always on the posterior end of the abdomen.

3. *State the seven main (compulsory or necessary) titles used in the classification of arthropods of medical importance.*
(i) Kingdom, (ii) Phylum, (iii) Class, (iv) Order, (v) Family, (vi) Genus, (vii) Species.

4. *State the main divisions of the above titles, which may be used or not, according to requirements.*
(i) Subkingdom, (ii) Subphylum, (iii) Subclass, (iv) Suborder, (v) Subfamily, (vi) Subgenus, (vii) Subspecies.

5. *What extra titles are there which may be inserted at almost any level?*
'Grade', 'Division', 'Series', 'Section', 'Group'.

6. *What are the last four letters composing the ending of all 'family' names?*
'-idae'.

7. *What are the last four letters composing the ending of all 'subfamily' names?*
'-inae'.

8. *What are the last three letters composing the ending of 'tribal' names?*
'-ini'.

9. *What are the last five letters composing the ending of a 'super-family'?*
'-oidea'.

10. *How many latinised words constitute a specific name? What is the name of the first of these words?*
Two latinised words, of which the first is the genus or generic title.

11. *How should a genus and species be written and printed?*
A genus should always begin with a capital initial letter; when the second specific part of the name is added, this should begin with a small initial letter. When written, a generic name and the specific name should always be underlined. This will indicate to a printer that the word or words are to be printed in 'italic' type (not 'roman' as the rest of the text).

12. *Compare briefly the characteristics of scorpions, spiders, mites and ticks.*
Scorpions are clearly segmented, well sclerotized elongate arachnids with small pincer-like chelicerae and large articulate pedipalps which are adapted for seizing prey. They have simple eyes on the prosoma shield. There are four pairs of walking legs to the prosoma and there is an opisthosoma composed of a segmented abdomen with a segmented tail, ending in a telson and poisonous sting. Their prey (*e.g.* grasshoppers) is crushed between the gnathobases of the pedipalps and fore-legs, and the body juices and pulp are sucked into the mouth opening.

Spiders are arachnids of more compact appearance than scorpions, but with a more impressive spread of legs. They show little trace of segmentation, though the scutum-covered prosoma is distinctly separated from the opisthosoma by means of a constricted waist. This is probably because all spiders have spinerettes on the end of the 'abdomen' and need to turn the opisthosoma about in order to weave the egg cocoon. Some species also weave web-snares.

They usually have eight simple eyes, though eyes may be absent. The pointed chelicerae contain a poison duct which opens at the cheliceral extremities. They also have gnathobases for crushing out the juices of their prey and use a digestive fluid to liquify the tissue within sclerotized portions such as femora. The pedipalps are leg-like (not pincers) and are smaller than the true legs.

Mites and ticks form the Order Acarina (*acari*—mite or tick) and are highly specialised arachnids with the unfed body usually compressed dorso-ventrally, and with no demaraction between the prosoma and opisthosoma. This gives the whole body a sac-like appearance. Paired chelicerae or jaws, and pedipalps, are always present, together with a large hypostome in ticks and a small one in mites. The great majority of acarina live on fluids from decaying organic matter or living plants and animals, the cell structures of which they sometimes break down outside their body by digestive enzymes, if the species concerned is not capable of direct ingestion.

The chelicerae and pedipalps are mounted on a base or capitulum. This is often called a false head, since eyes if present are never on this portion, but are situated further back on the anterior part of the body-sac.

The body integument may be thin and membranous (*e.g.* cheese-mites) or much thickened into plates and shields (*e.g.* some ticks).

There are no gnathobases. The legs are four in number in nymphs and adults, but the larva (or first stage out of the egg) has but three pairs (= slight metamorphosis). The anus is usually ventral and never at the posterior end of the body, though it is always posterior to the opening of the gonads.

13. *Describe* Pediculoides ventricosus *and make comments on its relation to disease.*
Pediculoides ventricosus, the straw itch or grain itch mite, belongs to the Family Tarsonemidae of the Superfamily Sarcoptoidea (which are without visible spiracles, and presumably breathe through the skin and mouth).

These mites inhabit straw, grain, and cotton seed, and feed on the body-fluids of insect pests such as grubs infesting grain.

Unfortunately the active females readily attach themselves to the skin of man and cause violent itching, which may be followed by urticarial rashes, rise in temperature, headache, nausea, or even vomiting and diarrhoea. Since mattresses filled with straw are sometimes heavily infested with starving female mites, the attack may be severe.

There is a marked sexual dimorphism. The male is light yellow, ovoid, flattened and just visible and has a shield-like appearance. The young female is elongately oval and looks rather like a grass seed when attached to the skin. The gravid female becomes enormously distended behind the fourth pair of legs with the development of the young mites, which actually hatch in the uterus and reach sexual maturity before leaving the body of the parent. A gravid female may give birth to nearly 300 adult young of which the great majority will be females. The life cycle under optimum conditions may take about one week. Dermatitis due to these mites has occurred in personnel handling grain and other cargoes in many parts of the world.

14. *Describe* Sarcoptes scabiei *and make comments on its relation to disease.*

Sarcoptes scabiei, the scabies mite, causes 'scabies' in man and 'sarcoptic mange' in domestic and some other animals. These mites are somewhat spherical in appearance (flattest below), and have the integument transversely wrinkled (reminiscent of 'finger-print whorls'). The dorsal surface has several rows of peculiar pedunculated spines or bristles (fourteen on posterior third of body) and two rows of cone-shaped bristles on about the middle of the dorsum (six, three on either side), together with a number of serrated projections arranged in transverse rows. A remarkable generic feature is the almost terminal position of the anus (rare among Acarina). The 'false head' which carries the short chelicerae and protective pedipalps somewhat resembles a turtle's head peeping from the body-shell. It carries two short bristles (a generic characteristic) on the dorsal aspect.

The legs of *Sarcoptes scabiei* are widely separated into two anterior pairs and two posterior pairs. They are short and stumpy, though of five segments, and end in a pair of short straight (compared to the curved single claw of *Tyroglyphus*) stout digging claws. Between these claws on the two pairs of fore-legs extends a long unsegmented pedicel or stalk on the extremity of which is a small sucker. A similar pedicel and sucker arrangement occurs on the fourth pair of legs of the male, while the third pair of legs of the male, and hind pairs of legs in the female bear long trailing bristles instead of the pedicel and sucker.

An infestation of scabies mites burrowing in the horny layers of the skin is not usually noticed until sensitisation has been acquired, which usually requires three weeks to a month. After this, intense itching occurs, and for years afterwards the patient will react to reinfestation by violent scratching within a few hours of the new contact.

15. *Describe* Tyroglyphus farinae *the common forage mite, and make comments on its relation to disease.*

Tyroglyphus farinae, the common flour mite, is a good example of a 'forage mite', of which several species exist on various stored food products. It belongs to the Family Tyroglyphidae and is (like the straw itch mite and the scabies mite) without spiracles, breathing through the cuticle.

It is a very small (just visible to the eye) somewhat pear-shaped and elongate, and usually of a pearly whiteness. The skin is smooth and not wrinkled transversely as in *Sarcoptes.* There are many long white bristles. The rounded posterior portion is separated from the anterior portion by a demarcation or waist, which also separates the areas on which are situated the pairs of legs. The front two pairs have their bases close together, and the two hinder pairs are similarly near to each other. They are all longish and structurally similar, and terminate in a single curved claw and sucker. The chelicerae are chelate or pincer-like and the palps are small. There are no eyes.

These mites may give rise to a troublesome allergic dermatitis if allowed to remain in numbers on human skin for any length of time, and may cause intestinal disturbance if swallowed, as in infested flour or meal.

There is a peculiar non-active nymphal form—the hypopus—which attaches itself to the body hairs of a housefly or mouse in order to be transported to new supplies of food.

16. *Describe* Trombicula deliensis *the 'Far Eastern' scrub-typhus mite, and make comments on its relation to disease.*

Trombicula deliensis, the 'Far Eastern' scrub typhus mite, is in the adult stage a blind, long-legged, distinctly waisted velvety red mite with diagnostically swollen front tarsi. Both the nymphs and adults feed on such material as insect eggs and small mites.

The orange coloured ectoparasitic larval stage is quite different in appearance to the nymphs and adults. It is very small (just visible to the eye), and ovoid. It has no waist and bears three pairs of legs only (nymphs and adults have four pairs). These legs are seven segmented and terminate in three long curved claws (two are true claws, the middle one is the empodium). No other arthropod of medical importance has this type of leg-ending.

The false head is distinct (as in *Sarco!tes*) and bears the scimitar-shaped pair of chelicerae and lateral pedipalps. On the anterior portion of the dorsum is the rectangular scutum or shield, which bears seven long pinnate bristles, each springing from a distinct pit or socket. On either side of the shield are the eyes, which resemble two small asterisks (on either side) when seen under the microscope. These enable the larva to crawl towards the sun when climbing a blade of grass or a twig in order to lie in wait for a passing host. There is a number of pinnate setae on the dorsum behind the shield which are of importance in identification.

These larvae after emergence from the eggs, climb to a suitable position and lie in wait for a passing host. On attachment, the larvae anchor themselves by means of their chelicerae and then form a histio-siphon in the host's skin by digesting away a tube outside their body. From this lesion they engorge themselves on the exudate. If their digestive saliva contains *Rickettsia orientalis*, then an eschar is likely to be formed at the site of the bite, and a human host may acquire scrub-typhus.

17. *Describe* Liponyssus bacoti *and make comments on its relation to disease.*

Liponyssus bacoti, the blood sucking 'tropical' rat mite, is a species of the family Dermanyssidae of the superfamily Parasitoidea of the suborder Mesostigmata. They are world wide ectoparasites of both the brown and black rat, and are the vectors of murine typhus in the southern United States of America.

These mites are small and elongately oval. The female body varies according to the degree of distension. There is an elongate shield on the dorsum shaped something like a blunt spear head which is proportionately larger in the male than the female. The

154

body has many shortish setae over its entire surface including the shield. The legs are comparatively longer than in *Tyroglyphus*. The mites feed entirely on blood at all stages and drop from the host after each meal. They readily attack man and sometimes swarm in warehouses and granaries. The bites may give rise to a severe dermatitis.

18. *What are Mesostigmata?*
Mesostigmata are a suborder of the Acarina, with a pair of spiracles medially placed on the lateral borders (*i.e.* on either side) of some mites and all ticks.

19. *Define the Superfamily Ixodoidea.*
The superfamily Ixodoidea or Ticks are medium to large meso-stigmate Acarina without a waist or visible metameric segmentation. They are all bloodsucking ectoparasites of vertebrate animals (amphibians, reptiles, birds and mammals). There is always a rostrum to which the mouth parts are attached, inserted into an anterior indentation. The mouth opening lies between the cheliceral sheaths above and the large projecting toothed hypostome below.

The ventral surface is always grooved. The integument is leathery and often ornate.

Ticks are world wide in distribution, but most abundant in the tropics and subtropics.

20. *Describe the Family Ixodidae.*
The Ixodidae or Hard Ticks are Ixodoidea which always have a dorsal shield—large in males and small in females—and therefore show a marked sexual dimorphism.

The capitulum or false head of which the mouth armature forms a part is always terminal and is clearly visible when viewed from above. The pedipalps are usually rigid though hinged about their base, and are hollowed out so as to act as sheaths to the chelicerae when these are not in use (palps do not enter the bite puncture). The hypostome is typically of uniform width and is armed with many rows of recurved teeth. The chelicerae are very strongly developed and protrude from cheliceral sheaths. A scutum or shield is always attached to the false head at the camerostoma and so to the main body of the tick. In the male it covers the dorsum and allows of little body distension. In the unfed female it occupies approximately the anterior third of the dorsum, but in the distended gravid female only a fraction of the body surface behind the capitulum is so covered. The eyes, when present, are dorsal and situated laterally on the scutum. Pulvilli are always present. The spiracles are usually large and are typically posterior to the fourth pair of coxae.

Hard ticks are typically 'permanent feeders', and moult twice only, being one-host, two-host or three-host parasites. They are mainly active during the day, and when the host is found, they remain attached for days or weeks. The eggs are laid in one batch on the ground, the female dying with her single mass. There is one larval and one nymphal stage. The males are much smaller than the gravid females and rarely feed. Five genera contain hard ticks of medical importance. They invariably require humid conditions in between periods on their host.

21. *Describe the Family Argasidae.*
The Argasidae or Soft Ticks are Ixodoidea which lack a dorsal shield or scutum, and show only slight sexual dimorphism.

The pedipalps are free and leg-like with their segments subequal in length. The hypostome is broader at the base than at the apex, with a smaller number of teeth than in the Ixodidae. Similarly, the chelicerae are more delicately armed than in hard ticks. The capitulum or false head to which the mouth parts are attached lies wholly, in nymphs and adults, in a hollow of the overhanging anterior hood (only partly so in larvae, the mouth parts of which appear more terminal). The mouth parts are therefore described as terminal or subterminal in larvae and ventral or 'not visible dorsally' in nymphs and adults. The integument has no shield, but may bear mammillae, tubercles or discs (according to species). The ventral surface (venter) shows a coxal fold (internal to the coxae), and a supracoxal fold (above the coxae). There is no great increase in length and breadth during engorgement (only a dorso-ventral swelling), or in pregnancy, compared to the enormous increase in size of gravid female hard ticks. Pulvilli are absent or rudimentary. Eyes, when present (*e.g. Ornithodorus savignyi*) are laterally situated on (above) the supracoxal folds. The spiracles are small, discoidal or reniform, and typically anterior to the fourth pair of coxae.

Soft ticks are typically rapid intermittent feeders (except the larvae of some species) and are often long lived (years). They are mainly nocturnal, feeding on sleeping hosts and returning to the dust or crevices in the habitat afterwards. These ticks usually prefer a dry habitat (hard ticks a humid one), and some species void large quantities of coxal fluid (separated blood plasma) in their chosen retreat, following a blood meal. Several batches of eggs are laid at intervals between feeding, but the total number does not equal that of the large egg mass of a female hard tick. There are several nymphal stages (only one larval and one nymphal in hard ticks).

22. *Comment on the relationship of* Ornithodorus moubata *to tick-borne relapsing fever.*
Ornithodorus moubata, the main vector of tick-borne relapsing fever, has two special reasons for its notoriety. Firstly, it is a domestic tick (also a pest of the caravan routes), and is largely dependent on man as host. Secondly, it produces coxal fluid more rapidly and in greater quantity than any other tick, so that if it is infected, the fluid will swarm with spirochaetes and readily infect the host through the open bite puncture. Since the spirochaetes invade all parts of the tick the ovaries become capable of trans-mitting the rickettsiae to the next generation. The saliva of an infected tick and the faeces may be of importance on occasion, but the coxal fluid is the main vehicle of transmission in *Ornithodorus moubata.*

All *Ornithodorus* appear to have natural spirochaetes, which are passaged through their normal wild hosts (rodents, goats etc.) without hurt to either. When man steps into this cycle, however, he may contract a fever which will vary in severity according to the 'species' of spirochaetes inoculated.

23. *How is the genus* Ornithodorus *recognised?*
The genus *Ornithodorus* is a genus of the Argasidae or soft ticks which have a flat body when unfed, but usually become very convex on distension. The anterior end may be pointed or hood-like. The lateral body-margin is thick and never clearly defined (the reverse is the case in *Argas*) and is of similar pattern to the rest of the integument so that all trace of a border disappears on engorgement (exception is *Ornithodorus dyeri*). The surface

157

integument is characterised by a tuberculated or mammillated appearance like that of shagreen. The eyes when present (absent in *Ornithodorus moubata* but present in *Ornithodorus savignyi*) appear as two pairs above the supracoxal fold. The tips of the pedipalps may be visible from the dorsal view.

Coxal glands open at the base of the second coxae and commonly discharge a considerable quantity of clear fluid (blood plasma) during engorgement. Species vary in size from about 3 mm. to 30 mm. The larval stages of some species vary from two to six or more. The hosts may be cold or warm blooded vertebrates.

24. *Comment on the relationship of* Dermacentor andersoni *to tick-borne typhus.*

Dermacentor andersoni is the Western North American hard tick mainly responsible for the form of tick-borne typhus known as Rocky Mountain Spotted fever. The causal organism is *Rickettsia rickettsi*. The disease is endemic in most of the United States, but the Rocky Mountain country is most heavily infested with the fever ticks. The reservoir appears to be the ticks themselves, which transmit the rickettsiae transovarialy to the following generations.

Infected ticks tend to congregate in certain areas, so that it may be dangerous to camp in one valley but be safe in the next.

The infection is acquired *via* the bite of the ticks, the saliva being the transmitting vehicle.

Tick-borne typhus occurs in many parts of the world and varies in severity and importance according to the vector species. For example, *Dermacentor andersoni*, the wood tick of Western North America is associated with an approximate mortality of 80%. *Dermacentor variabilis*, the dog tick of Eastern North America is associated with approximate mortality of 25%. *Amblyomma cajennense* of South America (Brazil) is associated with about 70% mortality, while *Rhipicephalus sanguineus* of the Mediterranean area causes about 20% deaths. Tick-bite fever of Southern Africa, conveyed from veldt animals to man via cattle and dogs is associated with about a 1% mortality.

25. *What is the meaning of the word 'Insect'?*

The word 'Insect' means 'cut into' (or divided into 'head', 'thorax' and 'abdomen').

158

26. *How many metameric segments are included in the head, thorax and abdomen of an adult insect, respectively?*
The head contains six segments which are indistinguishably fused to form a capsule. The thorax contains three usually discernable segments. The abdomen contains *not* more than eleven segments, of which from four (in houseflies and bluebottles) to eight (in mosquitoes and fleas) are usually easily visible.

27. *Name the typical appendages of an insectan head, thorax and abdomen.*
The insectan head bears (i) three pairs of mouth parts—pair mandibles, pair maxillae, and labium or lower lip (fused pair of second maxillae). (ii) One pair of antennae. (iii) Usually one pair of compound eyes (parasitic lice, fleas and some other insects are exceptions). Sometimes simple eyes or ocelli are present (on vertex of housefly and tsetse fly for instance).

The insectan thorax bears one pair of legs on the pro-thorax. One pair of legs, and usually one pair of wings (except in Apterygota and some parasitic insects such as fleas, lice, bed-bugs, and some other insects) on the meso-thorax. One pair of legs, and usually one pair of wings (except in Apterygota, some parasitic insects as fleas, lice and bed-bugs, and, of course, the Diptera where hind-wings have become halteres) on the meta-thorax.

The insectan abdomen has no appendages except the genitalia (examples are the paired cerci in female mosquitoes and the paired claspers on the hypopygium of male mosquitoes).

28. *What are Apterygote insects?*
The Subclass Apterygota (*a*, without; *pteron*, wing) are believed to be the oldest forms of insects. They are, like arachnids, without a tendency to develop wings. They typically have one or more paired abdominal appendages other than the external genitalia and cerci. Compound eyes and simple eyes may be present or absent. Typically there are distinct pro-, meso- and meta-thoracic segments. There may be eleven visible segments or less, which typically terminate in well developed cerci, suggesting a primitive condition. Another usual feature unknown in pterygote insects is that moults (ecdyses) in some forms continue after sexual maturity. Metamorphosis (change of shape during development from egg to adult) is slight or absent.

159

29. *What are Pterygote insects?*

The Pterygota are a subclass of insects which evolved from Apterygote stock and developed two pairs of flying wings on the meso- and meta-thorax respectively (there are never wings on the prothorax). The early forms distributed themselves widely due to their powers of flight and from them have evolved most of the different kinds of insects seen everywhere today. Living examples are grasshoppers, earwigs, dragonflies, may-flies, bugs, lice, butterflies and moths, beetles, flies and fleas.

30. *Differentiate between exopterygotes and endopterygotes.*

The differences between members of the divisions Exopterygota and Endopterygota, both of which belong to the winged Subclass Pterygota, though some parasitic forms are secondarily wingless, are due to their methods of developing from egg to adult.

Exopterygote insects tend to lay comparatively few large eggs from which hatch six-legged larvae resembling their parents in almost every way except for their small size, lack of flying wings, and sexual maturity. Typically the wings are externally visible as wing buds which increase in size with each moult.

Endopterygote insects tend to lay many small eggs on the larval food supply, from which hatch larvae with little resemblance to the adult forms. These immature stages eat voraciously and moult occasionally until they have gathered sufficient fat body, stored in globules, to undergo a non-feeding pupal instar, during which the sexually mature winged form is organised.

The exopterygote development is described as an incomplete metamorphosis (= hemi-metabolous), while the endopterygote development is described as a complete metamorphosis (= holometabolous).

There are also certain intermediate forms, which, however, do not invalidate these generalised divisions of the Pterygota.

31. *What is the order to which Bugs belong?*

The order to which Bugs belong is the Hemiptera. They are so named because a large proportion have fore-wings which are membranous and partly transparent at their distal extremity, and opaque or leathery at their proximal remainder, *i.e.* they are 'half winged'.

32. *How may a bug be distinguished from other insects?*
Bugs may be differentiated from other insects by their piercing and sucking proboscis (visible portion is the grooved labium bearing two pairs of piercing stylets), which is always segmented, and reflexed below the ventral surface of the head when not in use. The labium does not enter the host's tissue during the meal of plant-sap or blood (according to the kind of bug).

33. *How many wings do typical bugs have?*
Typical bugs are flying insects and have two pairs of wings (Diptera have but one). The hind pair of wings is always membraneous, and the fore-pair is the larger, stronger and thicker pair. Even water bugs fly from pond to pond, and the aphids have some generations apterous, and some with wings. In one Suborder, the Homoptera (the cicadas are well known examples) the fore-wings are typically of the same consistency throughout (often membranous), while the Heteroptera (to which the blood-sucking bugs belong), typically have the fore-wings mostly thickened with a distal portion thinner or even quite membranous. This is best seen in the common plant-sucking shield bugs.

34. *What are the two families which contain blood-sucking bugs of medical importance?*
The two families containing blood-sucking bugs of medical importance are the Families Reduviidae (which have flying wings in the adult forms) or Assassin bugs, and the *Cimicidae* (which never have flying wings) which contains the common bed bugs.

The family Reduviidae contains many predacious bugs which chiefly live by sucking the body fluids of other arthropods. One group, the Triatominae, however, has specialised in feeding on warm blooded animals. To this subfamily the cone-nose bugs of medical importance belong.

The family Cimicidae contains blood sucking ectoparasites of birds, bats and man. The two common species of bed-bugs (*Cimex lectularius* of Europe, North America and elsewhere, and *Cimex hemipterus* (= *rotundatus*) of southern Asia, Africa and elsewhere) belong to this family.

35. *Describe the appearance of common bed-bugs.*
The temperate region bed-bug *Cimex lectularius* and the tropical

region bed-bug *Cimex hemipterus* are similar in general structure as follows:

Bed-bugs are small (though larger than lice or fleas) being about a quarter of an inch long and almost as broad. Their colour is a uniform mahogany brown when examined under a low power microscope, though when caught in a room they may appear grey with dust, or even dark red from a recent blood meal.

The head is broad (as is the whole of the body) and carries a pair of projecting compound eyes, which are black in colour (fleas and lice have black simple eyes). There are no simple eyes (as in Triatominae). The front of the head projects beyond the antennal bases after the manner of a crab louse, *i.e.* it broadens anteriorly before rounding off. From this projection the proboscis or labium hinges forward from its recurved resting position when the bugs are about to feed. When at rest this apparatus folds tightly into a *groove* beneath the head of the bed-bug.

The antennae are of five segments and are comparatively long (lice have shortish five-segmented antennae and fleas have even shorter many segmented antennae, usually tucked away in a groove behind the eyes).

The upper plate of the prothorax is known as the pronotum and is characteristic in being markedly kidney shaped, the two ends of the reniform shield being the 'shoulders'. These shoulders are much more pinched out or cusped in the species *Cimex lectularius* than in *Cimex hemipterus*.

The mesothorax is seen as a small triangular wedge between the pronotum and the two hemi-elytra (half wing cases) which are all that is left of the ancestral pair of fore-wings (the hind wings have been completely lost). The wingless metathorax is seen as an even smaller wedge-shaped slip between the hemi-elytra and the first abdominal segment.

The legs are slim but strong (and longer than in lice), and the tarsal segments are three in number as in Triatominae (flies and fleas have five tarsal segments and lice one) and are used in a somewhat flat-footed manner to enable the bug to scuttle rather than to run. The legs terminate in paired claws as in flies and fleas (human lice have hinged single claws).

The abdomen is extremely broad and flat in an unfed bed-bug. It shows eight segments (nine in reality but the first is not easily seen). The female has a rounded abdominal extremity, while the male has a small indentation and hook-like projection pointing to the left. The female on the fourth visible sternite shows a remarkable nick in the lower edge known as the 'organ of Berlese'.

36. *Describe the appearance of American cone-nose bugs.*
The American cone-nose bugs of the subfamily Triatominae, which contains the many genera that cause American Trypanosomiasis in tropical America, are rather large bugs (three-quarters of an inch to one and half inches) of general similar structure as follows:

They are all elongate bugs (bed-bugs of the genus *Cimex* are remarkably broad) and as with bed-bugs, are very much flattened from above downward when unfed. When gorged with blood their abdomen distends greatly in an upwards and downwards direction but with little elongation (bed-bugs when engorged tend to lengthen).

The colouration varies from grey-brown (*Rhodnius*) to black, with red patches of colour on the thoracic projections and some bands of yellow and red on the abdomen which are partly obscured by the folded wings (*e.g. Triatoma*).

The head is long and cone shaped (tapering anteriorly). At its tip the proboscis is hinged, so as to allow the labium and stylets to swing forward to form a continuation of the long axis of the bug when feeding, and to hinge below the head when not in use. The proboscis does not touch the under surface of the head, however (bed-bugs have the labium resting in a groove below the head).

The antennae are longish and somewhat thin (five segmented and gymnocerate as in bed-bugs) and join the head some little way from the tip of the head. Behind the antennal bases is the pair of lateral projecting compound eyes (like 'blackberries', as in bed-bugs) and behind the compound eyes lie the pair of ocelli or simple eyes (absent in bed-bugs).

The upper plate of the prothorax, or pronotum, is somewhat wedge-shaped, and narrows anteriorly (bed-bugs have their pronotum kidney-shaped) and often bears small bumps or projections (not found on bed-bugs). The mesothorax shows as a small wedge shaped projection between the wing bases when the wings are folded on the abdomen.

163

The wings are four in number. The fore-pair is usually tough and leathery and strongly veined, and in some species is distinctly thinner at the distal end than the proximal (almost hemi-elytrous). The hind pair is always membranous and flimsy in appearance (bed-bugs have no wings, only a pair of vestiges). When not spread for flying the wings are folded flat on the back, with usually the lateral edges of some of the abdominal segments showing on either side.

The legs are strongly made but not nearly so stout as in lice, and the tarsal segments (as in bed-bugs) are three in number and are used in a flat footed manner during progression.

37. *How many immature stages are there in the life cycle of bed-bugs and cone-nose bugs?*
There are five immature stages between the egg and adult forms of bed-bugs and cone-nose bugs. Bed-bugs take about six weeks to produce a generation and cone-nose bugs about ten months. A bed-bug may live two months, a cone-nose bug (even when infected with trypanosomes) some two years.

38. *Describe the appearance of a bed-bug egg.*
The egg of the bed-bug (*Cimex lectularius*) is ovoid and somewhat sausage shaped, being slightly curved near its anterior end or operculum. It is a whitish pearly colour and about one millimeter long. Continuous polygonal patterns are usually visible over the surface of the egg, which appear white by top-light and are dark in a cleared specimen seen by the transmitted light of a microscope mirror. There are *no* pierced nodules on the operculum such as are found on louse eggs. There is a narrow rim to the operculum. Considerable larval development takes place while the egg is in the ovary. The normal period for hatching is four to eight days after oviposition.

39. *What disease is transmitted by cone-nose bugs of the subfamily Triatominae?*
American trypanosomiasis or Chagas' disease caused by *Trypanosoma cruzi*, is transmitted to man *via* the infected faeces of certain cone-nose bugs of the subfamily Triatominae. The blood sucking cone-nose bugs habitually defaecate whilst feeding on their host (bed-bugs usually defaecate in their resting place behind wallpaper or in cracks in the furniture), and infective metacyclic

trypanosomes may swarm in the dark liquid waste deposits. The trypanosomes gain access to the human host through the bite puncture, through the mucous membranes of the mouth or nose, or through the conjunctiva of the eyes.

40. *Are lice insects?*
Lice belong to the Class Insecta because they have their body segmentation divided into a head, thorax and abdomen. They have one pair of antennae, and three pairs of legs, and the opening to the gonads is at the posterior end of the body.

They have *no* compound eyes, however, being either blind or with a pair of simple eyes. They have *no* wings either, as these would not suit their ectoparasitic habits.

41. *Though wingless lice belong to the Subclass Pterygota and to the division Exopterygota. Make comments.*
Lice are small parasites which infest the skin of animals and birds. They are descended from cockroach-like insects and therefore belong to the Exopterygota. This means they hatch from largish eggs in the form of miniature adults. There are three immature stages, all of which (except for size) resemble closely the final adult forms. Lice are secondarily wingless due to their parasitic life, though their ancestors developed wings in buds outside the thorax, before this specialisation by reduction evolved.

42. *How are lice recognised, and what is their distribution?*
Lice belong to the Order Anoplura. They are small wingless ectoparasites of birds and mammals (*i.e.* warm blooded hosts as with fleas. *Glossina palpalis*, it will be remembered will feed on cold blooded hosts, as will some mites and ticks.). They are flattened dorso-ventrally and have a characteristic tough partly sclerotized chitinous integument. The legs are short and stout, and their endings are specially adapted for clinging to (in some species gripping) the hairs or body of the host. The distribution of lice is world wide, wherever hosts are available to give them suitable shelter, and they are invariably host specific—even having strong predilictions for certain parts of the host—the head of man (*Pediculus humanus capitis*) and neck feathers of some birds being examples (Mallophaga spp.). While fleas have strong host preferences, they are able to change from one warm blooded host to

165

another in time of need. Lice, however, die after a few hours or days away from their specific hosts.

43. *Make a brief comparison to differentiate between the Suborders Siphunculata (blood sucking lice) and Mallophaga (cuticle gnawing lice).*
The suborder Mallophaga, are chiefly bird lice with some species infesting mammals. (*Trichodectes canis*, is a mallophagous louse which sometimes ingests the eggs of the dog tapeworm. If the louse is accidentally swallowed by a human, usually a pet-loving child, he or she may develop the worms). The head of these lice is broader than long and usually broader than the thorax. There are strong mandibulate mouth parts for chewing at dead particles of skin or feathers. The prothorax is typically free and distinct.

The suborder Siphunculata, which are mammalian lice only, typically have the head narrower than the thorax. Their mouth parts are *not* mandibulate, but have become adapted for piercing and sucking. They are also retractable. The thorax always has its three segments fused (the prothorax is separated from the other two segments in Mallophaga). The tarsi are never more than one segment and the claws are always single (some Mallophaga have the normal double claws).

44. *On what hosts are members of the Family Pediculidae found?*
Blood sucking lice belonging to the family *Pediculidae* are found only on lemurs, monkeys, apes and man, *i.e.* the primates. The pair of simple eyes is always pigmented and prominent (some Siphunculata are without eyes).

45. *On what hosts are species of the genera* Pediculus *and* Phthirus *found?*
Lice belonging to the genera *Pediculus* and *Phthirus* are found on the higher apes and man only. They are host specific.

46. *Differentiate between* Pediculus humanus *and* Phthirus pubis.
Pediculus humanus, the 'common louse', is an elongate active louse (compared to *Phthirus pubis*), which like *Phthirus pubis* is found on man only. This louse contains two varieties which have a distinct preference for the part of the host on which they feed. The smaller, and it is believed the original form, spends its life on the hairs of the human head. The other variety (*Pediculus humanus var.*

corporis) the larger, and later developed form, spends its life on the body parts other than the head, hands and feet. After feeding they tend to wander about the clothing and when on the outer garments are easily passed from one person to another.

These two varieties are able to interbreed and therefore are difficult to divide into separate species, and in the following description are treated as one.

Pediculus humanus are small (up to $\frac{1}{8}$ inch). The general colour is grey, and they are often soiled with the blood faeces of their fellows. Occasionally an over-feed of blood will rupture the midgut, and by colouring the haemocoele give a bright red colour to the louse. Such a louse becomes very active but soon dies. The head of *Pediculus humanus* is not broader than long, and carries a pair of shortish five segmented antennae (if the antennae are long then the insect is not a human louse, even if the rest of it resembles one).

The front of the head between the antennae is somewhat pointed and carries the small retractable mouth parts. The eyes are darkly pigmented and prominent on either side of the head. The thorax, flattened like the rest of the louse from above downward, is somewhat wedge-shaped (with its narrowing taking place forward). Though the three segments of the thorax may be traced yet they are not distinct and are fused into one strong base for the three pairs of legs.

The legs are characteristic, no other arthropod of medical importance (except *Phthirus pubis*, the crab louse) being so strongly designed and armed with single gripping claws on a single tarsal segment, opposed to a thumb-like projection on the tibia.

This unique apparatus enables the louse to not only cling firmly to body hairs or clothing fibres, but to ambulate comfortably on body or clothes in a way impossible to insects constructed to run on smooth surfaces.

In *Pediculus humanus* all three pairs of legs are of approximately equal development, with the exception of the fore-legs of the males, which have the claw endings more strongly made than in any of the other legs.

There are no wings or traces of wings. The elongate abdomen, which has a clear demarcation where it joins the thorax, is widest at about the middle of its length and is remarkable for a series of indentations along the lateral aspect. On the convex portions between these indentations are the spiracles, which are on or very near the lateral edge of the abdomen in the centre of dark sclerotized shields.

Phthirus pubis, the pubic-hair or crab louse, by comparison with *Pediculus humanus* is squat and broad, adopting by its sedentary habits the appearance of a tiny tick rather than an elongate active insect. These lice spend most of their lives on one spot with their mouth parts in the skin of their human host, sucking blood when they require a meal. They therefore have a body shape and structure adapted for filling with blood (and in the female to contain the large exopterygote eggs) while their powerful claws cling firmly to the host, in spite of scratching efforts to dislodge them.

The head of the crab louse is similar to that of body and head lice in having similar eyes and antennae, but the front of the head from the bases of the antennae to the snout instead of tapering to a blunt point actually enlarges.

The thorax is much wider than long and is so compressed and fused with the abdomen as to obliterate any sort of hinged waist, and, indeed, any clear demarcation between it and the abdomen.

The forelegs are characteristically weak compared to the forelegs of *Pediculus humanus* and bear long thin single claws. To make up for this lack of robustness, the second and third pair of legs of *Phthirus pubis* have far and away the most powerful hair-gripping apparatus of any insect of medical importance, for, not only are the hinged claws of large size but the tarsus and tibia are fused together to make a strong base for the claws to grip against. There is a thumb-like process on the tibia which opposes the claws. All the leg segments are shorter and broader than in head and body lice.

The abdomen is quite different to that of *Pediculus humanus*. The spiracles are not lateral in the crab louse, but internal to the margin and not surrounded by dark sclerotized plates. Of the six abdominal spiracles visible the upper three have been forced into a horizontal row by the compression of the abdomen into the thorax. This is diagnostic.

The abdomen tapers somewhat from its base to the posterior end, which is bifid in the female and rounded in the male. Crab lice, however, possess four pairs of lateral processes on the abdomen which are easily seen from above, and which terminate in longish hairs. The wide bifid appearance, caused by the posterior projections in the male crab louse must not be confused with the true narrow bifid cleft of the female.

47. *Describe the eggs of human lice?*
The eggs of *Pediculus humanus* and *Phthirus pubis* are similar in being largish ovals deposited on a blob of liquid louse-cement, which is first applied to a hair or clothing fibre. This rapid-hardening clear-cement is diagnostic, as bed-bugs and other human insect parasites produce no such adhesive mass.

A further characteristic of louse eggs is the operculum or rimmed lid which is forced off by the emerging larval louse. Bed-bugs and cone-nose bug eggs also possess an operculum, but these two parasites lack the pierced nodules on the lid of the operculum of human lice.

48. *How many immature stages of lice are there?*
There are three immature stages of *Pediculus humanus* and *Phthirus pubis*. The life cycle proceeds as follows: egg stage; first nymphal or larval stage; moult; second nymphal or nymphal stage; moult; third nymphal or mature nymphal stage; moult; imago or adult louse.

49. *What diseases are conveyed by* Pediculus humanus*?*
Pediculus humanus is the vector of classical epidemic typhus. The organisms causing this disease (*Rickettsia prowazeki*) are transmitted from infected man to man by the faeces of infected lice. The organisms tend to kill the louse within a week to two weeks.

Louse-borne relapsing fever is caused by *Borrelia recurrentis* and transmitted from infected man to man in the ruptured body fluid of *Pediculus humanus*. The spirochaetes do not of themselves harm the louse.

50. *What is the food of human lice?*
The food of *Pediculus humanus* and *Phthirus pubis* is human blood. All stages, except the egg, feed readily and pass blood faeces on the host.

51. *To what Order do fleas belong?*
Fleas are placed in the order Siphonaptera Latreille 1825, though Karl Jordan suggested that Suctoria Retzius 1783 had priority.

Continental entomologists tend to use the order Aphaniptera Kirby 1817 (though this was published without a valid description).

52. *What are the chief morphological characteristic of fleas?*
Fleas are always small ($\frac{1}{32}$ inch to $\frac{1}{8}$ inch). They are always brown in colour (light chestnut to almost black, according to species). They are also markedly compressed laterally, and are thus the only insects of medical importance to be so flattened. Their exterior is completely sclerotized by chitinous plates which typically fit over each other in a beautifully streamlined manner (often assisted by thin flanges and rows of strengthening bristles). The female Jigger flea is exceptional in showing a large amount of intersegmental membrane when gravid. The legs have remarkably large coxae and are adapted for running and jumping, particularly the hind legs. Fleas never have wings.

53. *What niche in nature do fleas occupy?*
Fleas are of world-wide distribution, wherever their animal hosts live under conditions suitable to their breeding habits. Mammals and birds regularly occupying nests or litter are favourite hosts of fleas. Cattle, horses, sheep etc. do *not* harbour fleas, but rats and mice, domestic fowls, dogs, cats, humans etc. are parasitized whenever possible.

54. *Are fleas endopterygote or exopterygote?*
Fleas belong to the Endopterygota, because they show a complete metamorphosis in their life history, including a pupal stage. They are therefore closely related to the higher *winged* insects, but due to their specialised parasitic life have secondarily lost their ancestral wings.

55. *Give a simple classification which divides the fleas of medical importance.*
The fleas of medical importance fall simply into two categories: (1) 'Free-running fleas' or those that habitually move about on the host's body (though they may spend long periods in the nest material or bedding), and (2) 'Stick-tight fleas', which in the female at least, anchor themselves permanently in one position in the host's tissue and remain there while the egg mass develops.

56. *Give some genera of the groups referred to in the question above. (55).*
Examples of 'free running fleas' are *Pulex* (there is one species only in this genus, *Pulex irritans* the 'human' flea); *Xenopsylla* (including perhaps sixty species such as *Xenopsylla cheopis*, the plague or tropical rat flea, *Xenopsylla astia*, another tropical rat flea, and *Xenopsylla brasiliensis* a mainly African rat flea); *Nosopsyllus fasciatus* (the temperate region rat flea, believed responsible for the Black Death and Great Plague). *Ctenocephalides felis* and *canis*, the cat and dog fleas respectively; and *Leptopsylla segnis*, the mouse flea.

Examples of 'Stick-tight fleas' are *Echidnophaga gallinacea*, the tropical fowl-flea, and *Tunga penetrans* the tropical African and American Jigger flea.

7. *How is a male flea distinguished from a female?*
The female flea typically presents an oval posterior outline and in an unfed or 'cleared' flea the dark spermatheca should show up distinctly. The male flea typically has a pointed posterior, and in an unfed or 'cleared' flea the many whip-like lines of the male organ are seen to occupy the major part of the abdomen.

58. *What are 'combs'?*
'Combs' on fleas are distinctly flattened bristles, usually arranged in a row along the edge of a segment or leg. They invariably point backward and are important in identification. For instance neither the 'Plague flea' (*Xenopsylla cheopis*) nor the 'Human flea' (*Pulex irritans*) have combs, while the common temperate region rat flea (*Nosopsyllus fasciatus*) and the cat and dog fleas (*Ctenocephalides* (= comb head) *canis* and *felis*) always have combs on the prothorax. The 'combs' do not have a pit or socket at their bases, but spring direct from the integument. Combs may consist of but one or two projections, or a row of many teeth. They are usually black in colour.

59. *What is the life history of a typical flea?*
The life history of a typical flea starts with eggs laid by the gravid female when on the host. These pearly white eggs (which resemble castor sugar grains when seen on a dark background) hatch into small maggot-like larvae, which seek organic debris for their food

171

supply. Since adult fleas tend to indulge in orgasms of blood sucking, the excess blood is voided on the hairs of the host during the act of feeding. These blood clots on drying find their way into the sleeping quarters of the host, or the ground. The larvae feed voraciously by means of their mandibulate mouth parts on such material, and, after moulting three times, pupate within a silken cocoon, to which much camouflaging debris adheres. The adult flea, when ready to emerge, may postpone its exit from the micro-climate of the cocoon until such time as the stimulus of the passing footfalls of a host cause it to emerge suddenly and leap towards this passing source of food. The larvae of bird fleas are usually found in their nests, those of rodent fleas in their sleeping and breeding quarters, and those of man in between floorboards or dusty skirtings, or even in some cases on the bodies of unclean persons in amongst the clothes.

60. *Why is* Xenopsylla cheopis *considered the most dangerous of all fleas?*

Xenopsylla cheopis is considered the most dangerous of all fleas because it is the best vector of *Yeisinia pestis*, the causal organisms of bubonic plague, and of *Rickettsia mooseri*, the causal organisms of murine typhus. *X. cheopis* appears to have a proventricular valve which 'blocks' with plague bacilli more readily than most other fleas. It also seems to be more resistant to adverse conditions of heat and humidity than other tropical fleas, and to be a more persistent biter when starving under the conditions of a 'blocked' or 'partly-blocked' proventriculus, than its relatives.

In murine typhus the causal *Rickettsiae* are passed from rat to rat by its specific louse *Polyplax spinulosus*. This little louse does not bite man. The tropical rat flea *X. cheopis*, however, does bite man if it becomes separated from its rat host, and, if it has acquired the pathogenic organisms from an infected rat, it will, on occasions, pass on the infection.

61. *How are flea larvae to be differentiated from sandfly larvae?*

Flea larvae are whitish active little maggots of some thirteen segments (behind the head capsule). There is a tapering off anteriorly, and an increase of size towards the posterior segments, which, however, suddenly appear to diminish in size. Such a little maggot could easily be confused with either a calypterate fly larva or a sandfly larva. There are, however, two characters which are not found in either of these fly larvae—long body hairs (the sandfly larvae have 'club-shaped' short body hairs, and the calypterate fly larvae have partial body rings bearing shortish spines) and ANAL STRUTS. These are formed of two backwardly pointing sclerotized projections at the extreme end of the flea larva's body, which point downward. They and the body hairs are used in progression.

Sandfly larvae lack anal struts, but they do have CAUDAL BRISTLES, which point upwards—two in the first stage larva, and four in the second, third and fourth stages.

Both flea larvae and sandfly larvae have a head capsule bearing short antennae and mandibulate mouth parts, and both are blind, spending their lives buried in the darkness of their environment and food material. The flea larva has only three stages to the sandfly larva's four.

62. *Do adult fleas leave their cocoon as soon as they are fully developed?*
Fleas pupate within a silken cocoon to which adheres sufficient local debris to camouflage the whole. Inside this cocoon the thin-walled pupa (with the last larval skin shed at its posterior end, but *not* adhering to its tail segments as in the *Phlebotomus* pupa) rapidly develops into an adult flea. This adult, however, does not usually emerge until the vibration of a passing host stimulates it to push aside the silken strands with its head and to try and jump on the host. Fleas may remain inside the microclimate of the cocoon for months before emerging to seek a blood meal.

Fleas may, however, emerge and breed without a blood meal, laying a small number of viable eggs which will help to build up a population in untenanted premises where there is a good larval food supply (suitable organic material).

63. *Describe briefly the mouth parts of a flea.*
The mouth parts of a flea are composed of a labrum epipharynx, a pair of mandibles, a pair of maxillae and their palps, and a labrum with a pair of longish palps. The hypopharynx which is short and dagger-like terminates about the bases of the mandibles and is the outlet of the salivary duct.

173

When a flea begins to feed on its host it cuts a puncture with the sharp blade-like mandibles. Neither the labrum epipharynx nor the maxillae, nor any of the palps directly assist in this wound, which is made in order to suck blood straight into the pharynx.

64. *How would you recognise a* Xenopsylla *flea?*
Xenopsylla of medical importance are typically lightly pigmented medium length fleas with a prominently dark simple eye on either side of the head, and with a clearly marked vertical internal reinforcement between the mid-thoracic sternite and mid-thoracic pleuron known as the 'mesothoracic bar' (more correctly the mesopleural suture). The lower end of this 'bar' connects with the coxa of the second leg. No *Xenopsylla* species is blind (as is the mouse flea *Leptopsylla segnis*), nor do they ever display combs (as in the rat flea *Nosopsyllus fasciatus*), and none has a tooth on the genal angle (as in the rat flea *Pariodontis*). They are well made, compact little fleas, neither markedly elongate (as in the common chicken flea *Ceratophyllus gallinae*) nor noticeably ornamented with bristles (as in the blind *Hystrichopsylla* the mole flea, which, by the way, has two spermathecae). The mesopleuron (with the 'bar') is wider than in the human flea, *Pulex irritans* (which though it contains a mesosternal transverse thickening present also in *Xenopsylla*, lacks the 'vertical bar' of the mesopleural suture so characteristic of *Xenopsylla*).

In *Xenopsylla* there is a row of bristles along the posterior border of the head (where the neck would be if present). In *Pulex* there is but one bristle in this position. In *Xenopsylla* the antepygidial bristles are well developed while in *Pulex* they are shorter and much less prominent.

65. *How would you recognise a* Xenopsylla cheopis?
A *Xenopsylla cheopis* female (with the oval posterior outline) has a spermatheca in which the transparent portion of the tail is well curved over the head, reminding one of the letter C.

A *Xenopsylla cheopis* male (with the pointed posterior outline) has the 'ninth sternite' quite unlike the eighth or preceding sternites, in being 'club shaped' and vertically situated within the eighth enclosing segment, only its head or tip being above this eighth segment.

66. *What is a Dipterous fly?*
A dipterous fly is an insect with one pair of wings and one pair of halteres. It undergoes a complete metamorphosis including a pupal stage, and has mouth parts which can imbibe fluid food only.

67. *What is a Nematocerous fly?*
A nematocerous fly is a form of dipteran which has long thread-like antennae of not less than eight segments. Six of these must be essentially similar, though the two basal (scale and pedicel) may be differently shaped.

68. *What is a Culicid fly?*
A culicid fly is a 'gnat-like' rather than 'midge-like' fly which is classified in the family Culicidae. Examples are the Dixinae, Chaoborinae and Culicinae. Some older authors refer to mosquitoe as Culicidae. It is not important on this course to differentiate culicids as such.

69. *What is a member of the Culicinae?*
A member of the subfamily Culicinae is a mosquito. It differs from other flies in having a characteristic wing-venation and a long projecting proboscis and projecting palps. Other Nematocera likely to be mistaken for mosquitoes have the proboscis pointing ventrally and the palps recurved below the head *e.g.* sandflies.

70. *What is the characteristic venation of a mosquito?*
The wing of a mosquito is long and narrow, and rounded at the apex. The third vein is short and simple and lies at the tip of the wing. It is situated between two forked veins. All six veins and the costal margin have many tiny flat scales attached to them and there is a marked fringe of these scales along the hind border of the wing.

71. *Name and explain the position of the mouth parts of a female mosquito.*
The mouth parts of the female mosquito comprise two parts. Firstly, the food channel, which is composed of a dorsal tunnel made up of the much extended roof of the mouth or labrum epipharynx, and supported ventrally by the flatter extension of the mouth floor or hypopharynx (which also houses a part of the long salivary duct throughout its length). Secondly, to assist the insertion of this sharp ended food tube into the host's tissue, there

175

is a pair of long extremely thin mandibles with splayed cutting tips lying alongside the labrum epipharynx; and also, more importantly, there is a pair of similarly long thin maxillae with backwardly toothed splayed endings, which lie close to the hypopharynx.

These two parts, made up of six individual tools are always together as a fascicle or bundle, which when at rest lies in a gutter in the dorsal portion of the much larger labium or lower lips. This labium sheath never enters the bite puncture, but always retains its hold on the fascicle at its tip by means of its labella.

72. *Describe and name the main structures of a mosquito thorax.*
The large dorsal cover is the meso-notum or scutum. Behind the wing bases it is strengthened by a transverse raised belt or scutellum, which, at its lateral extremities joins the upper corners of the sterno-pleuron, which in turn curves around and supports the ventral aspect of the thorax. From the anterior ventral part of the sterno-pleuron to the anterior part of the scutum reaches the small forked pro-thorax. The large membranous area between sterno-pleuron and prothorax supports the anterior spiracle and the posterior pronotum. Posterior to the sides of the sterno-pleuron are seen the walls of the mesepimeron and meron, which are continued by the lateral walls of the metathorax (which close the rear of the thorax and carry the halteres and posterior spiracles). The meta-thorax is roofed by the heart-shaped post-notum, which joins the scutellum to the first abdominal segment.

73. *Describe very simply the visible parts of a mosquito abdomen.*
A mosquito abdomen appears to consist of eight segments. The plates on the dorsal surface are nota (singular notum) or tergites (singular tergum). Those on the ventral surface are sterna or sternites (singular sternum). The elastic semi-transparent lateral portion is known as the pleuron. It supports the abdominal breathing spiracles.

The eighth segment appears to give rise to a pair of finger-like cerci in the female, and to a pair of claspers (with the coxites, styles and claspettes as the most recognisable parts) in the male.

74. *Describe briefly the alimentary tract of a female mosquito.*
The piercing and sucking mouth parts of a mosquito are followed

176

by the pharynx and oesophagus, which in the thorax joins the beginning of the long cardia or neck of the mid-gut by way of a stout valve known as the proventriculus. Just prior to the proventriculus there branch dorsally two bladder-like dorsal diverticula, and ventrally one longer and larger ventral diverticulum.

The ventriculus or stomach or mid-gut is the largest vessel in the body and is found in the abdomen.

It terminates at the juncture of the five malpighian tubes or renal organs, which mark the connection of the mid-gut to the ileo-colon portion of the hind-gut. This leads to the rectum which contains the six conical rectal papillae which conserve water by re-absorption from the excreta if necessary.

75. *Describe an Anopheline mosquito.*
An anopheline mosquito has no bands of scales on the abdominal segments. The scutellum has a simple posterior outline and a regular row of bristle-pits and backwardly pointing bristles. The palps of the male are as long as the proboscis and 'clubbed'. The palps of the female are as long as the proboscis and 'not clubbed'.

76. *Describe a Culicine mosquito.*
A culicine mosquito has transverse patches of light and dark overlapping scales on the abdominal segments giving it a distinctly 'banded appearance'. The scutellum has a tri-lobed posterior border, and the bristle-pits and their backwardly projecting bristles are gathered on to the three lobes. The palps of the female are much shorter than the proboscis and the palps of the male are long and variously terminated, but never typically 'clubbed' as in an anopheline male.

77. *What is the observable difference between the abdominal terminations of females of the genera* Culex *and* Aedes?
The eighth segment of a female *Culex* is broad and the cerci not very long. This gives a blunt appearance. The eighth segment of an *Aedes* female is rather narrow and the cerci are comparatively long. This gives a pointed appearance (particularly when the eighth segment protrudes during abdominal engorgement and distention).

78. *Describe briefly* 'Anopheles-*type*' *eggs.*
'*Anopheles*-type' eggs are boat-shaped. They are laid singly in

batches, on water, and typically have a pair of lateral air-float chambers. The shell is thin and easily dessicated.

79. *Describe briefly* 'Aedes-*type' eggs.*
'*Aedes*-type' eggs are ovoid. They are laid singly in batches, typically above the water line, or in situations liable to be flooded. They have no lateral float chambers, and the shell is resistant to adverse conditions.

80. *Describe briefly* 'Culex-*type' eggs.*
'*Culex*-type' eggs are typically club-shaped. They are laid together on water to form rafts of a hundred or so eggs. There are no float chambers, though the rafts float well. The shells are thin and the larvae hatch downward into the water.

81. *Draw the feeding position of Anopheline larvae.*
Anopheline larvae typically feed in a position horizontal with or parallel to and just below the water surface.

82. *Draw the feeding positions of Culicine larvae.*
Culicine larvae typically feed in semi-vertical positions as they hang from their tail siphon from the water surface, or swim to the bottom to chew at sunken debris.

83. *Tabulate briefly the characteristics which differentiate Anopheline larvae from Culicine larvae.*
Anopheline larvae have heads typically longer than broad (which they rotate when feeding).
The lateral balancing bristles on thorax and abdomen are mostly pinnate or feathered.
They have a pair of retractile notched organs on the prothorax (which assists floating).
They have paired palmate float hairs on the dorsum of most of the abdominal segments.
Median tergal plates are present.
The spiracular plate which protects the pair of breathing spiracles is 'flush' with the eighth segment.
The anal segment typically has two pairs of anchoring bristles.

84. *Tabulate briefly the characteristics which differentiate Culicine larvae from Anopheline larvae.*

The Culicine larva has a head typically broader than long (which it swims with, rather than rotates).

The lateral balancing bristles on thorax and abdomen are never pinnate or feathered though they may branch.

There are no tretractile notched organs, palmate float hairs, tergal plates or anchoring bristles.

There is, however, a large projecting siphon on the eighth segment which protects the spiracles and breathing tracheae and which at once differentiates the larva from an Anopheline.

85. *Draw a mosquito pupa, with its organs of respiration and apparatus for swimming.*

A mosquito pupa is 'comma-shaped', with a large cephalo-thorax, and a flexible segmented abdomen ending in a pair of swimming paddles. It breathes through a pair of 'trumpets' which allow the anterior spiracles of the developing adult to obtain atmospheric air.

86. *What is a sandfly?*

A sandfly is a biting Psychodid, of the Suborder Nematocera of the Order Diptera. (This is not the kind of question likely to be asked in an examination, but it is the way an entomologist thinks of an insect.)

87. *What is the most noticeable thing about a sandfly when it flies near the observer?*

The dark compound eyes are most noticeable. If the sandfly is of a buff colour, as most species are, this dark moving spot is often all the observer can focus on.

88. *In what two ways do the nematocerous antennae of sandflies differ from the nematocerous antennae of mosquitoes?*

The antennae of sandflies are closely covered with short hairs so as to hide the segmentation. The antennae of mosquitoes have a whorl of hair to each segment, leaving the individual segments visible.

The antennae of sandflies are similar in both sexes. In mosquitoes the males have bushy (plumose) antennae, the females sparsely haired (pilose) antennae.

89. *Name five forms of disease typically associated with the bites of certain sandfly species.*

179

(1) Sandfly fever, (2) Oriental sore or cutaneous leishmaniasis, (3) Kala azar or visceral leishmaniasis, (4) Mucocutaneous leishmaniasis or espundia, (5) Oroya fever or bartonellosis or Carrion's disease. The bites of sandflies may cause a severe reaction, especially in newcomers to sandfly areas.

90. *Describe the appearance of a* Phlebotomus *or sandfly.*

The *Phlebotomus* or sandfly adult may be described as follows:

Size—is very small, less than 3 mm., and will pass through 18 mesh screening and mosquito netting.

General colour—is grey or golden yellow to dark brown.

General appearance—is very hairy, with the back humped, and the wings normally are held at an angle of 45° above body.

Head—proboscis short (adapted for biting and sucking blood in females) and projects ventrally (mosquitoes have a porrect proboscis).

Palps—are of five segments, and are recurved below head (mosquitoes have porrect palps).

Eyes—usually noticeably dark in colour.

Antennae—are long and threadlike, of sixteen segments, and are similar in either sex, though relatively longer in males. (*Psychoda* antennae are 'beaded'.)

Thorax—legs are very long and slender (*Psychoda* has shorter legs), wings are long and narrow, pointed at the tip, and the second longitudinal vein forks twice, once at about the middle of the wing length and again nearer the upper portion of the tip. Halteres are relatively large.

Abdomen—may have erect or recumbent hairs on terga according to species.

Genitalia—male terminalia or genital claspers are typically large and conspicuous, especially the upper pair. Female has pair of shortish cerci.

91. *What is the life history of sandflies?*

The life history of sandflies (Phlebotominae) begins with the tiny eggs being laid in damp crevices away from the light. There are four grub-like larval stages which occur below ground in darkness. These are followed by a quiescent non-feeding stage with a remarkably transparent skin, which eventually gives rise to the winged adult. The life cycle is about two months and the adults typically live for one week to three weeks. The females feed some two to four days after emergence from the pupae.

92. *Describe how you would recognise :*
  (i) a first stage sandfly larva,
  (ii) a second, third or fourth stage larva,
  (iii) a sandfly pupa.
(i) First stage larva.
Size—is tiny.
Colour—has whitish grey body with dark head.
Appearance—is elongate and grub-like, and has no eyes, but has pair of short mandibulate mouth parts.
Has three thoracic segments lacking ventral false legs, and ten abdominal segments bearing a single pseudopod each.
Body hairs are somewhat club-shaped and typically have globular terminations.
Has *two* upwardly projecting caudal bristles almost as long as the body on tail segment. These are called caudal bristles.
(ii) The second, third and fourth stage larvae are similar in appearance to the first stage, except that they are progressively larger in size and carry *four* upwardly projecting caudal bristles on the tail segment. These caudal bristles are not so long relatively as in the first stage.
(iii) The pupa is somewhat transparent and displays the developing adult through its cuticle. The legs are almost free in their projecting sheath-like processes. The fourth moulted larval skin is invariably retained in a crumpled mass on the caudal extremity. This skin will show the diagnostic club-shaped body hairs and four long caudal bristles. The pupa has a concave-shaped dorsal aspect and shows a series of short 'spinal' projections.

93. *What are the habits of sandflies.*
Phlebotomus adults are mainly crepuscular or nocturnal, hiding in dark, warm damp places by day. The females are usually indiscriminate blood feeders with a preference for bovines. Sandflies will take sweet vegetable fluid at times.

Sandflies progress by short hops when not flying. They do not fly strongly, and avoid winds and draughts. Usually found within 50 yards of their breeding places, unless wind-blown. Females will pass through mosquito netting to feed, but usually remain inside the net when gorged with blood.

94. *What forms of Filariasis are transmitted by insects other than mosquitoes?*
Dipetalonema (= *Acanthocheilonema*) *perstans*, and *Mansonella ozzardi* by *Culicoides* biting midges; *Onchocerca volvulus* by *Simulium* biting black-flies; and *Loa loa* by *Chrysops* gad flies.

95. *At what time of day and under what conditions are these vectors of filariasis most likely to bite?*
Culicoides bite most readily at dusk, and may continue throughout the night. (They frequently creep amongst the hairs on the head and limbs in order to bite.) The females attack singly or in swarms.

Simulium bite during the day, preferably in windless warm overcast conditions such as before a thunderstorm. They do not bite during darkness but may do so by moonlight. The females attack singly or in swarms. There are three habitual types. Those that attack birds; those that prefer mammals; and those that will feed on man as well as other mammals.

Chrysops are most aggressive in warm sunshine linked with a humid atmosphere. They are less active during dark cloudy weather or cool rainy periods.

96. *What type of terrain is usually associated with these vectors?*
Culicoides are particularly associated with marshes and damp ground. Heaps of rotting leaves under trees are the breeding grounds of some species. Attacks by swarms may be serious, rendering some areas difficult to work in outdoors. Examples are parts of Florida, Atlantic seaboard, southern states of North America, West Indies, western Scotland. Bites are sharply painful and the irritation may last several days.

Simulium are particularly associated with rapidly flowing rivers and waterfalls, or shallow clear water, the essential requirement being well oxygenated water for the immature stages. Some species breed on the wave-splashed rocky shores of some African lakes. The bites are not usually painful, but there may be very severe reaction and swelling of body parts such as the eyelids.

Chrysops are particularly associated with woodland or scrub where their host animals exist, and which is reasonably near marshy ground, where the larvae mostly feed on rotting vegetation. The adult population is usually associated with suitably wet conditions during the previous year as the life cycle typically takes several months. The bite causes immediate pain, and local swelling and reaction may be severe.

97. *Compare the antennae of these vectors. Mention the Suborder to which they belong.*
The nematocerous antennae of *Culicoides* are of 15 segments, the majority of which are shorter or more bead-like than in mosquitoes and sandflies. They are, like those of mosquitoes, plumose in males and pilose in females. The second segment or pedicel as in mosquitoes and sandflies is swollen and of larger size than any in the flagellum.

The nematocerous antennae of *Simulium* are of 9–11 segments in different species. They are similar in both sexes (as in sandflies) and have no swollen pedicel, which, however, is distinct from the scape (thus differing from mosquitoes, sandflies and *Culicoides*). They are tightly beaded (*i.e.* touching) and taper at the extremity. There are no whorls of hair.

The brachycerous antennae of *Chrysops* are of fewer segments than are found in Nematocera (which have at least eight segments). In this genus, where the antennae are particularly long and projecting forward (as compared to most other Tabanidae, such as *Tabanus*) the first three segments are seen to be well separated from each other with each segment larger than the preceding. The third terminates in a spiky style of four distinct annulations, which is slightly upturned in life. These antennae are always longer than the head (front to rear).

98. *How can the sex of these vectors be simply ascertained?*
In *Culicoides* the antennae are bushy in the males and sparsely haired in the female.

In *Simulium* the antennae are similar in both sexes but the males have joining or holoptic compound eyes while the females have separate or dichoptic compound eyes.

In *Chrysops* the antennae of both sexes are similar, but the male eyes are noticeably holoptic and the females noticeably dichoptic. Both sexes have three simple eyes or ocelli on the vertex of the head (which though found on the heads of house flies, tsetse and their like, are never found on the heads of mosquitoes, sandflies, *Culicoides* or *Simulium*).

99. *What is the comparative size of these vectors?*
Culicoides species are very approximately 1 to 2 mm.; *Simulium* 2–5 mm.; and *Chrysops* (small Tabanids) about 5–10 mm., *i.e* very small; small; and about the size of housefly.

100. *What is different in the function of the mouth parts of two of these vectors to the remaining one?*
The mouth parts of the female *Culicoides* are somewhat similar to a shortened version of the mouth parts of a mosquito, except that, like *Simulium* their mandibles function on a scissors principle.

The mouth parts of the female *Simulium* have shear-like mandible blades that snip the host's skin like scissors, with a pair of maxillae that tear open the cut tissue, and a food channel of labrum epipharynx, mandibles and hypopharynx that is inserted into the blood which is released. The sheath-like labium with the pair of horny labella does not enter the wound.

The mouth parts of the female *Chrysops* are like a more powerful and specialised version of the *Simulium* type, but lack the scissors principle. There is a pair of large sabre-like cutting mandibles, a similar pair of maxillae and two large segmented distinctly visible palps which do not enter the bite puncture. The usual sucking channel is present. A large labium with labella is easily visible.

101. *Describe the wings of these vectors.*
The wing of *Culicoides* is short, broad and rounded at the tip. It is very simply veined and bears two enclosed cells at the distal end of the radial or first longitudinal vein. These are sometimes a little difficult to see clearly. Most *Culicoides* have spotted wings due to pigment of the membrane (spotting in mosquito wings is always due to clumps of wing scales) and their many microtrichia (microscopic short hairs on the membrane) and macrotrichia (larger hairs on the wing veins and margins) are of importance in identification.

The wing of *Simulium* is very deep from the fore to the hind margin, and noticeably membranous with a venation that appears to be largely made up of foldings. The few thick veins are close to the radius or first longitudinal vein and are all well forward near the

184

anterior edge. The large area of membrane often gives rise to pale rainbow-like iridescence. The wing tip is rounded.

The wing of *Chrysops* always has dark bands superimposed on the typical wing venation of the family Tabanidae. This venation has some three enclosed cells in the centre of the wing, giving rise to radiating veins that make a number of open cells at the outer and posterior margins. The wing shape is designed for powerful flight and has a somewhat pointed tip.

102. *Describe the leg endings of these vectors.*
Culicoides has small paired claws on the fifth tarsal lag segment.

Simulium has a similar but somewhat stronger pair of claws as leg terminations.

Chrysops not only has the terminal pair of stout sharp claws, but has a fleshy pad beneath each, known as a pulvillus, together with a central pulvilliform empodium. These organs help the fly to get a silent grip on the slippery hairs of the host animal.

103. *Mention the method of breathing of two of these vectors in the larval stage, and how do these vary from the breathing arrangement of the remaining larval form?*
The larvae of *Culicoides* are mainly aquatic and extract oxygen from the water in which they live (or in some species from the damp air of their habitat) by means of anal gills—processes on the last abdominal segment.

The larvae of *Simulium* are entirely aquatic (save for some forms that live on spray-covered rocks and the like) and extract oxygen from the water in which they live through retractile rectal blood gills on the last abdominal segment. *Simulium* larvae moult six times (mosquito larvae four times).

The larvae of *Chrysops* bears an attenuated siphon on the eighth abdominal segment (somewhat after the manner of a Culicine larva). The larvae are usually found in mud and have six moults.

104. *Compare very briefly the pupae of these vectors and state in what environment they are to be found.*
The tiny pupa of *Culicoides* is brown in colour, and typically hangs (floats) vertically (cephalothorax uppermost) in the water from the surface film. It has a pair of long breathing tubes with small

tubercles on each. The abdomen is mobile and typically squirms or rocks. There are no paddles at its extremity.

The pupae of *Simulium* are mainly found in fast running streams, inside an open ended silken cocoon. The closed end of this structure always points up stream. The breathing tubes (variously branched) of the pupa project from the open end of the cocoon. The developing adult within the pupa obtains oxygen from the water through these processes.

The larger pupae of *Chrysops* are typically found just below the surface of drier ground near the larval breeding ground. They are composed of a head, thorax and abdomen and rather resemble a butterfly chrysalis. Breathing spiracles can be seen along the sides of the broad flattish abdomen. The development of the head portion is mainly due to the large compound eyes of the developing adult. There is a kidney-shaped pair of spiracles lying between the head and thorax (these do not project as in mosquito pupae).

105. *What occurs when the skin forms of* Onchocerca volvulus *are ingested by a vector species of black-fly?*
The skin forms (microfilariae) of *Onchocerca volvulus* become active and pass through the mid-gut wall into the haemocoele and thence to the thoracic muscles. Here they develop and moult several times before passing to the head and labium of the *Simulium*. They reach the human host via the labella of the labium during a subsequent blood meal by the vector.

106. *Is the pattern of transmission of the filarial worms referred to in Question 94 similar to that given in Answer 105.*
The pattern of transmission of filarial worms is always the same. Ingestion is followed by migration to the thoracic muscles and later exit to the host is achieved *via* the labium of the vector.

107. *Define the characteristics of the Suborder Athericera.*
The Athericera, which are specialised 'higher flies', have very short antennae, consisting of one small stem against the face; a second larger segment (which has a split in its outer aspect in the schizophorous Athericera); and a much larger and broader third segment which bears on its dorsal surface an arista or long bristle (comprised of the three remaining segments of the flagellum), as well as sensory organs. These antennae should be compared with

the antennae of Brachycera (*e.g. Chrysops*) and Nematocera (*e.g. Anopheles*).

108. *Why are Schizophora so named?*
Schizophora are athericerous Diptera which carry an inverted U-shaped dark scar on the front of the head above and around the antennal base and hollows. This scar marks the site of the withdrawn ptilinum or bladder with which this series of flies forces off the pupal cap (cyclorrhaphous method) and push their way to the surface through earth particles or debris. On emergence to the atmospheric air the flies withdraw the membranous ptilinum and
The Aschiza are athericerous Diptera which have no ptilinum or ptilinal suture.

109. *What is the transverse suture?*
The transverse suture is a groove which crosses the dorsum of the thorax of calypterate flies (an atypical member is *Fannia canicularis*, the lesser house-fly which not only has a poor transverse suture but has squames so small as not to cover the halteres from above). This mesonotal groove is situated half way between the neck and the anterior edge of the scutellum.

110. *How many abdominal segments are usually visible from above in calypterate flies?*
The usual number of abdominal segments seen from above in calypterate flies is four. Examples are the house-flies, stable-flies, all the calypterates of medical importance shows a greater number, namely seven (six are comparatively easy to count). In *Auchmeromyia luteola*, the Congo floor-maggot fly, the second visible segment is much longer than the others.

111. *Describe the wing and calypter (squame) of a calypterate fly.*
The wing of a calypterate fly is always membranous and without scales and of roughly triangular shape with a rounded apex and anal angle.
It has two systems of veins to support its surface. The anterior portion contains the costa, the subcosta, the radius or first longitudinal and the second and third longitudinal veins. The posterior portion contains a fan-like pattern of the fourth, fifth and sixth

longitudinal veins. These are joined together by a few cross veins and the vein systems of the two portions are linked through a small cross vein and fourth vein which curves upwards to join the costa near the wing tip.

There are two lobes at the proximal posterior indentation of the wing—the alula and the wing-squame. On the side of the thorax there is a thoracic squame, which in typical calypterate flies covers the halteres from above. This covering (calypter or squama) gives the name to the series.

112. What are pulvilli?
Pulvilli are the adhesive pads found at the tip of the feet of a number of small arthropods. In calypterate flies there are two pulvilli, one below each of the paired claws (in Tabanidae there appear to be three to each foot).

113. What is the 'vomit-drop'?
The 'vomit drop' is the bubble of liquid often seen at the tip of a non-biting calypterate fly's proboscis between the labella. Its source of liquid is the crop or abdominal diverticulum of the fly, and its primary purpose is to keep the delicate membranes of the sucking surface of the lips moist. If the crop is filled with liquid containing pathogenic organisms, these may easily be transferred to human food *via* the vomit drop or the wet surfaces of the labella.

114. What is meant by 'mechanical transmission of pathogenic organisms via the body hairs'?
The 'mechanical transmission of pathogenic organisms *via* the body-hairs' occurs when calypterate flies (the housefly *Musca domestica* is the most notorious) pick up infected material and dust on their numerous body hairs and bristles, as well as the leg hairs and pulvilli.

As flies are continually combing themselves with the leg hairs, particles attached to the above mentioned parts are continually being deposited on the surfaces where the flies settle to clean themselves. This may cause a mechanical (as opposed to incubational or developmental) transmission of organisms from infected sources to human food or the person.

115. In what way is the faecal deposit of calypterate flies important in the spread of the excremental diseases?

Calypterate flies may defaecate some of the contents of previous liquid meals whilst feeding and thus contaminate human food. Defaecation takes place much less often than the application of the vomit-drop to food material. Dysentery bacilli remain alive within the house-fly for at least four days, and multiplication of some pathogenic organisms takes place in the fly's gut before being passed out, usually in the form of dark thick spots.

116. *Do female calypterate flies only produce eggs?*
Some kinds of female calypterate flies, such as house-flies and bluebottles, lay batches of whitish eggs; some other (but fewer) species produce first stage larvae (sometimes eggs that hatch immediately) such as grey-flesh-flies (*Sarcophaga, Wohlfahrtia*); while the tsetse (*Glossina*) is remarkable in depositing one fully grown larva (third stage) every ten days or so.

117. *What is characteristic about a calypterate fly larva?*
Though variously specialised to suit different food environments, calypterate fly larvae display a constant morphological pattern. They have no head capsule as do the larvae of mosquitoes, sandflies, biting midges and blackflies, but instead have a softish pointed vestigial cephalic termination with a subterminal mouth-opening from which inveraiably protrude the points of a pair of strong darkly pigmented mouth-hooks (cephalo-pharyngeal skeleton), used for holding or tearing open the cells in their food supply in order to liberate organic fluid. They also use the hooks during progression. The body walls of the larvae are invariably flexible and have partly formed rings of small or large spines (according to species) which aid the larva in penetration of its typical decaying vegetable or animal food. The posterior end is more or less truncate and bears the pair of spiracular plates—somewhat circular brown objects bearing slits of different patterns in different species, which 'designs' are important in identification. The outer circumference is called the peritreme, the spiracular openings the slits, and the small enclosed spot, the button.

118. *What is the typical food of calypterate flies of medical importance?*
Non-biting calypterate flies, such as houseflies and stable flies, breed mainly in fermenting vegetable matter, though dead animal

material is also readily accepted. Other non-biting calypterate flies such as bluebottles, green-bottles and grey flesh-flies much prefer carcasses to any other living food supply, and occasionally lay their eggs or deposit their first stage larvae on wounds or sores (= semi-specific myiasis). Others still, have larvae which develop only in living tissue (= specific myiasis). Examples are some African yellow-bottles. Finally one genus (*Glossina*) rears its maggots on internal secretions within the female abdomen up to the fully mature stage.

119. *Name and briefly define the three principal kinds of myiasis.*
Myiasis or the invasion of living tissue is of three principal kinds as follows:

(a) Specific myiasis, when fly larvae can develop *only* in or on living tissue, *e.g.* the African Tumbu yellow-bottle (*Cordylobia anthropophaga*).

(b) Semi-specific myiasis, when fly larvae invade the living tissue of wounds or sores. They tend to attack the necrotic tissue first, but if present in numbers (single larvae invariably die) and if the patient has not strong resistance (as in old people) then the larval enzymes may destroy the area around the wounds or sores, and cause much damage. This kind of myiasis is particularly serious when the nasal and aural passages are invaded.

(c) Accidental myiasis, when fly eggs, larvae or pupae are accidentally ingested in food material such as unwashed greenstuffs, and live for a short or long period in the human alimentary tract, where they may cause intestinal disturbance. The infestation of the uro-genital tract is also included in this heading.

120. *What is a puparium?*
A puparium is the name given to calypterate fly pupae, which characteristically have the last (third) larval skin, left as an outer protection to the pupating third stage larvae.

When ready to pupate this larva shrinks in length and becomes yellowish (as opposed to the normal whitish colour of earlier stages). As pupation proceeds the yellow colour deepens to light brown and finally to deep brown or even black in some species. When this dark colour has appeared the period of pupation has taken over, and the miraculous process of lysis and reconstruction within the true pupal cuticle is proceeding rapidly. The puparium, therefore, is applied to those flies which have the last larval skin hardened around the pupal cuticle (= coarctic pupation) as opposed to the pupae of many other insects, such as mosquitoes and fleas, which cast off the last larval skin and so have a more transparent unprotected pupal cuticle (obtectate pupation).

121. *What characteristics are typical of the Order Diptera?*
The Order Diptera consists of insects which have one pair of wings, one pair of halteres, mouth parts adapted for sucking fluid food only, and a complete metamorphosis.

122. *What characteristics are typical of the Suborder Athericera?*
The Suborder Athericera consists of flies which have short bulbous antennae of three segments, the third being much the largest and bearing a dorsal arista or spine (consisting of the three remaining segments of the flagellum).

123. *What characteristics are typical of the Series Schizophora?*
The series Schizophora consists of athericerous flies which have a large dark inverted horse-shoe shaped scar on the front of the head which marks the line of withdrawal of the bladder-like ptilinum, an organ used during emergence from the puparium and while pushing up through covering matter to reach the freedom of the atmospheric air.

124. *What characteristics are typical of the Section Calypteratae?*
The section Calypteratae consists of schizophorous flies which typically have a cleft on the outer side of the second antennal segment, a transverse mesonotal suture, a characteristic wing venation and a pair of thoracic wing squames that usually cover the halteres.

125. *Name the twelve medically important non-biting genera of calypterate flies, and the two medically important biting genera.*
The medically important calypterate flies are contained in some twelve non-biting genera, and some two biting genera.

The non-biting genera are *Musca, Muscina, Fannia, Calliphora, Cochliomyia, Chrysomyia, Lucilia, Phormia, Cordylobia, Auchmeromyia, Sarcophaga, Wohlfahrtia.*

The biting genera are *Stomoxys* and *Glossina.*

There are many other genera of non-biting and biting Calypterate flies of medical interest, but the genera listed contain the most important species.

126. *How are these genera grouped into families and subfamilies?*
These genera of Calypteratae of medical importance are grouped for convenience into two families and four subfamilies.

Musca, Muscina, Stomoxys, Glossina and *Fannia* belong to the Family Muscidae.

Calliphora, Cochliomyia, Chrysomyia, Lucilia, Phormia, Cordylobia, Auchmeromyia, Sarcophaga and *Wohlfahrtia* belong to the Family Calliphoridae (which has a row of hypopleural bristles below the posterior spiracles lacking in the Muscidae).

Musca, Muscina, Stomoxys and *Glossina* are grouped into the Subfamily Muscinae, because of similarities. *Fannia*, being atypical and more closely related to many other dissimilar outdoor flies is placed in the Subfamily Anthomyinae.

Calliphora, Cochliomyia, Chrysomyia, Lucilia, Phormia, Cordylobia and *Auchmeromyia*, which are closely related are grouped into the Subfamily Calliphorinae, while the grey flesh-flies *Sarcophaga* and *Wohlfahrtia* are separately placed in the Subfamily Sarcophaginae.

127. *Describe the genus* Musca.
The genus *Musca*. While there are several species in this genus, *Musca domestica*, the house-fly of world wide distribution is by far the most notorious. It is the chief carrier among insects of the organisms causing typhoid, dysentery, cholera, poliomyelitis etc.

The larvae have on occasion been incriminated in accidental and semi-specific myiasis (they can easily be reared on a meat diet), but they are not considered to be myiasis type maggots and breed mostly in fermenting vegetable matter.

The adult flies have a number of characteristics useful when comparing them with other related flies.

The arista (spine formed of 4th, 5th, and 6th vestigial antennal segments) is plumed with spinulae on both the upper and lower aspects, which is the usual 'ciliate' form met with in the non-biting calypterates of medical importance (*e.g. Calliphora*, the bluebottle). *Fannia* and *Wohlfahrtia* both have 'bare' aristae, and the two biting muscids *Stomoxys* and *Glossina* have 'fimbriate' aristae, *i.e.* are plumed with spinulae on the upper side only.

Musca domestica is a predominently dark grey coloured fly, inclining to black when viewed against a light background. A closer examination, however, reveals a certain amount of orange and yellow in the abdomen, particularly in the males. The thorax has four characteristic dark stripes, with paler areas between. Viewed from a different angle, the dark stripes become light and the paler areas dark. This change of pattern, due to the surface structure of the fly in relation to the angle of the light falling on it, is present to some extent on all calypterate flies, and is perhaps most marked on the abdomen of *Sarcophaga*, one of the grey flesh flies group. Here the typical 'chequer-board' pattern on the dorsal surface, of silvery squares alternating with darker squares reverses its pattern dramatically as the observer changes the position of his eyes.

The wing of *Musca domestica* is the standard to which the wings of other calypterates are compared. It shows the fourth longitudinal vein gently bending up almost to meet the third vein at the costal margin just above the wing tip.

The thoracic squame is well developed, being large and waxy-white and completely covers the halteres from above.

The compound eyes of the male house-fly, though not touching, are much nearer together than are those of the female, which are widely separated. (In the blue and green-bottle flies the eyes of the males are holoptic, as indeed, are most of the calypterates of medical importance—the most important exceptions being *Musca domestica* and *Glossina spp.*)

128. *Describe the genus* Muscina.
The genus *Muscina*. The commonest fly of this genus, often mistaken for *Musca domestica* is *Muscina stabulans*, the non-biting stable-fly (*Stomoxys calcitrans* is the biting stable-fly).

This fly is a little larger and stouter than the common house-fly and is a little lighter in its greeny-grey colouration.

The aristae are plumed on both sides as in most non-biting calypterates of medical importance. The eyes of the male are very close, and almost holoptic. In the female they are widely separated.

There are four dark stripes on the thorax, but the two lateral ones are indistinct. The scutellum is rather pointed and of 'rubbed' appearance.

The wing venation is quite distinct from any *Musca*. The fourth vein does not 'bend' but instead curves up to meet the costal vein about the wing tip. It does not closely approach the third vein as in *Musca* and so leaves the first posterior cell 'widely open'.

This venation is very similar to that of the biting fly *Stomoxys*, but there can be no confusion if it is remembered that *Muscina* has a retractable non-biting sucking proboscis, while *Stomoxys* has a non-retractile forwardly projecting dark brown shiny proboscis, as well as kidney-shaped eyes when seen from a lateral position (*Muscina* has normal D-shaped eyes when seen from the side).

Muscina is most attracted by rotting vegetable garbage. It does not fly around in rooms to anything like the extent that house-flies and lesser house-flies do. It does visit human excreta and also enters houses and settles on food. It is not, however, a myiasis type fly.

129. *Describe the genus* Fannia.

The genus *Fannia*. There is any number of outdoor flies to be found with the characteristic wing venation of the lesser house-fly, which is the best known species of this genus.

This venation is that the fourth vein runs almost parallel with the third vein (*i.e.* without marked 'bend' or 'curve') until both end at the costal vein, the third at about the wing tip and the fourth well below it, leaving the first posterior cell 'very widely open'. This is very different to either *Musca* or *Muscina* (or *Stomoxys* or indeed others in our list of calypterates) and stamps any flies with this type of wing habitually found in domestic premises as *Fannia*.

There are two common indoor species—*Fannia canicularis*, the numerous lesser house-fly, so called because it is smaller than *Musca domestica*, and the less numerous and darker *Fannia scalaris*, the latrine fly.

The aristae of *Fannia* are 'bare' (so are the aristae of *Wohlfahrtia*, which, however, has an entirely different wing venation, the fourth vein being sharply angulate).

The male eyes are closely approximate while the female eyes are widely separated. The somewhat elongate male abdomen contains yellow transparent areas not present in the heart-shaped gren-grey abdomen of the female. The majority of *Fannia canicularis* found in houses are males, which take little interest in food, but seem to fly and settle intermittently on any hanging object such as an electric light pendant.

They, however, do walk about on human food from time to time, and the females will hunt out any animal excreta (such as a cat's earth tray) in a house in order to lay their eggs.

The larvae have numerous fleshy processes extending from their segments, which enable them to survive in semi-liquid faeces which would drown the more orthodox type of calypterate fly larvae.

Fannia larvae have been known to cause accidental myiasis and intestinal disturbances. As they are the only larvae which can develop in urine-soaked rags, they have caused many cases of uro-genital myiasis.

130. *Describe the genus* Calliphora.

The genus *Calliphora*. These are the temperate region bluebottle flies. They are rather large, very stout, bristly flies, and the gravid females are determined seekers of dead animal matter, including meat, raw or cooked, intended for human consumption. They enter houses in search of meat or fish more readily than any other calypterate flies. Occasionally, but not typically, the larvae cause semi-specific myiasis and accidental myiasis, but they are not myiasis type flies. They are more concerned with the spread of the excremental diseases.

The colour of *Calliphora erythrocephala* and *Calliphora vomitoria* of this country is mainly metallic blue, with dark bands if observed closely, and with a peculiar dusty appearance which is characteristic of the genus.

The compound eyes of the males are very close together, but those of the females are widely separated.

The wing venation is similar to that of *Musca* save that the bend in the fourth vein is almost pointed, and the vein itself is distinctly separated from the third as they both join the costal vein above the wing tip. The squames of *Calliphora* are distinctive, however, having a pale edge or surround, and a dark closely haired central portion. The aristae are plumed on both sides.

The abdomen has rather the same proportions as *Cordylobia*, that is, not elongate but almost of semi-circular outline. This is of use in separating *Calliphora* from *Sarcophaga*, and *Cordylobia* from *Auchmeromyia*.

131. *Describe the genera* Cochliomyia *and* Chrysomyia.
The genera *Cochliomyia* and *Chrysomyia*. These contain the tropical region bluebottles. *Cochliomyia* comprise the New World species, and *Chrysomyia* the Old World species. In general colouration they have a brightly burnished blue, greenish-blue, or purplish-blue background with, however, in most cases distinctly dark bands on thorax and abdomen.

They may, however, be distinguished from all the other calypterate flies of medical importance by the absence of long stout bristles in rows on the dorsum of the thorax.

This is particularly useful in differentiating them from *Calliphora* bluebottles, and bluish specimens of *Lucilia* green-bottles.

The mesonotum or dorsum of the thorax is further covered in fine hairs in *Cochliomyia* and *Chrysomyia*.

Their aristae, as in all the Calliphorinae, are plumed on both sides. The cheeks are usually orange or buff, and the squames are typically hairy (*Lucilia* has waxy white bare squames).

Some of these flies have larvae which cause serious specific and semi-specific myiasis.

132. *Describe the genus* Lucilia.
The genus *Lucilia*. These comprise the temperate region and tropical green-bottle flies, which, however, often display a greenish or purple shade.

They are never dark banded, however, and always show well developed longitudinal rows of black bristles on the thorax (dorso-central and acrostichal bristles) which differentiate them from *Cochliomyia* and *Chrysomyia*. The squames are waxy white and without hairs on the upper surface and may be seen clearly from several feet distance.

Some species, or perhaps strains of some species, show a predilection for ovipositing on living animals such as sheep, and under certain conditions in human wounds or sores. The larvae are able to penetrate living flesh if present in numbers, and dissolve

196

healthy tissue so as to feed on it. The 'sheep maggot' flies of the British Isles and Australia are mainly of this genus.

The wing venation is similar to *Calliphora* with a rather sharp angle in the fourth vein.

133. *Describe the genus* Phormia.

The genus *Phormia*. These flies are usually of burnished black colouration though there are green and blue species. There are no simple diagnostic characteristics, and the contained species are not of great medical importance. *Phormia regina* is a well known warm weather species in Europe. Its larvae have been known to cause semi-specific myiasis.

134. *Describe the genus* Cordylobia.

The genus *Cordylobia*. The important species of this genus is the tropical African *Cordylobia anthropophaga*, a matt yellow or testaceous (*testa*, a shell) calliphorine, the female of which lays eggs in situations smelling of human (or animal) usage. Areas used for urination or for drying clothes on the ground are favourite places for oviposition. The young larvae or maggots on hatching endeavour to attach to a human being (or animal) and if successful, burrow superficially into the host's tissue. Here they remain until ready to pupate (moulting *in situ* twice), causing a boil-like lesion to occur. There is always a danger of secondary infection in these infestations.

The larvae may easily be identified by their breeding site and by the pattern of their posterior spiracles, but the adults have little to differentiate them from several other testaceous calypterates. If, however, larvae are removed from boil-like lesions and hatch into yellow-bottle flies it is almost certain that they will be *Cordylobia*. These maggots by the way give the best defined picture of specific myiasis.

This fly cannot be confused with *Auchmeromyia* because there is no elongate second visible abdominal segment.

135. *Describe the genus* Auchmeromyia.

The genus *Auchmeromyia*. This genus contains one species of medical importance—the Congo floor-maggot fly, *Auchmeromyia luteola*. The larvae live in the dust and crevices of African native huts and emerge at night (after the manner of soft ticks or bed-bugs) to suck blood from persons sleeping on the floor. They attach themselves by their mouth hooks and sucking mouth-opening and do not retire until engorged with the blood of their host. Large numbers of larvae feeding frequently cause lassitude and malaise.

The adult flies, though testaceous yellow as in *Cordylobia*, are easily distinguished from any other calypterate by the pointed abdomen and the extraordinarily long second visible abdominal segment.

They, like *Cordylobia* (and *Glossina*), only inhabit tropical Africa.

136. *Describe the genus* Sarcophaga.

The genus *Sarcophaga*. This is one of the larvipositing genera, as the females invariably deposit first stage larvae or eggs that hatch immediately, on their food supply. They will breed in both decomposing vegetable or animal material, and their big powerful larvae have sometimes been incriminated in causing wound myiasis as well as accidental intestinal disturbance.

The adult flies are a metallic grey colour, with red eyes, and are rather elongate (compared to *Calliphora* bluebottles). They are noticeably bristly and have large waxy white squames, and large pulvilli. The thorax is strongly striped with black and silver lines and the abdomen markedly chequered in black and silver 'squares'.

These flies are very powerful for their size and are strong fliers. They occasionally enter houses in search of food, when they may become carriers of the excremental diseases.

They are found in both temperate and tropical regions.

137. *Describe the genus* Wohlfahrtia.

The genus *Wohlfahrtia*. The females of this genus, like *Sarcophaga*, are larvipositors of first stage maggots, and some species (*e.g. Wohlfahrtia vigil* of Canada) have been known to deposit their young on the unbroken skin of children and to cause myiasis in this way. They normally breed in decomposing vegetable or animal material, but they may cause semi-specific myiasis and accidental intestinal disturbance.

The adult flies are similar in build to *Sarcophaga*, but the aristae are bare (as in *Fannia*) and the abdomen instead of being chequered in changing squares of black and silver has a matt grey background with rows of rounded black spots on its dorsal surface.

These flies are all very bristly. *Wohlfahrtia magnifica*, the spotted flesh fly of Southern Europe, Asia and North Africa, is a known cause of animal and human myiasis. The large size of the maggots of Sarcophaginae always renders an infestation by them dangerous, especially if near the eyes, ears or brain.

138. *What two genera of calypterate flies are concerned with the transmission of African trypanosomiasis?*
The two genera of calypterate flies concerned with the transmission of African trypanosomiasis are *Stomoxys* and *Glossina*. These two genera are able to transmit by mechanical means the trypanosomes on their wet proboscis up to 4-8 hours. This occurs during 'interrupted feeding'. Other flies such as Tabanids (Brachycera) may be similarly implicated, but in any case the 'mechanical transmission' is not nearly as important as the 'cyclical or incubational transmission' of *Glossina* species. This occurs after a tsetse's initial blood meal on an infected human or animal, when for a period of up to three weeks the parasites multiply within the *Glossina* alimentary tract. These eventually issue *via* the saliva as metacyclic infective trypanosomes able to cause the disease after inoculation into a suitable host during the vector's blood meal.

139. *Describe briefly the size, colour, arista, proboscis, compound eyes, wing venation and 'angle of rest', squames and abdomen of Stomoxys.*
Stomoxys—Size—about ¼″ or the size of a common house-fly.
Colour—dark greenish-grey.
Arista—poorly plumed on the upper side only.
Proboscis—shiny dark brown tube with bulbous base hinged to
    rostrum on lower part of head. Slightly 'swamped' or broadened
    at its distal end where small 'teeth' exist to tear at the skin of the
    host and cause blood to appear. This is sucked up by the fly.
    The proboscis hinges vertically downwards in its entirety when
    Stomoxys is feeding and returns to its porrect position after-
    wards.
Compound eyes—close together in males, widely separated in
    females. Of kidney shaped outline when viewed from the side
    (*Glossina* and *Muscina*, the two other genera most likely to be
    confused with *Stomoxys* have normal D-shaped eyes when
    viewed laterally).
Wing-venation—typical of a calypterate fly with a normal shaped
    discal cell (*Glossina* has a 'butcher's cleaver' discal cell), but the

fourth vein curves to the costal vein at about the wing tip almost in the same way as in *Muscina*, leaving the first posterior cell 'widely open'. This is quite unlike *Glossina* or *Musca*.

When at rest on a vertical surface, *Stomoxys* invariably has the head uppermost (*Musca domestica* has the head invariably downwards), and the wings are folded over the abdomen at a wider angle than in the 'common house-fly'.

Squames—not very large and do not always cover the halteres.

Abdomen—shows the typical number of segments of a calypterate fly, *i.e.* four (*Glossina* has seven, six being easily visible).

140. *Describe briefly the above listed parts of* Glossina.

Glossina

Size—$\frac{1}{4}''$ to $\frac{1}{2}''$ according to the species (*Glossina tachinoides* being small and neat, while *Glossina fusca* could be described as large and robust).

Colour—always a brown of sorts, but with much variation according to species.

Arista—plumed on the upper side only with beautifully radiating regular spinulae. These rays are further barbed on either side, thus making the arista of *Glossina* easy to differentiate from any other calypterate fly.

Proboscis—always projecting forward from underside of head whether the fly is feeding or at rest. The visible portion is composed of the hollowed palps. The food channel is mainly composed of labrum epipharynx, hypopharynx and labium, and descends from the protection of these palps during the act of feeding, by hinging vertically from its bulb-like base so as to pierce the host's skin below the vector's head. It is plunged into the tissues by a series of probing movements which apparently cause horizontal pools of blood to extravasate. These, affected by anticoagulant saliva, enable the tsetse to distend itself with blood in a short time.

Compound eyes—are wide apart in both males and females (as in *Musca domestica*) but when compared, the male eyes are closer together than those of the female. Are of normal D-shape when observed from the side (*Stomoxys* are reniform).

Wing-venation—rather longer and narrower than most calypterate fly wings (tsetse are powerful and rapid flyers) and the fourth vein joins the costal vein near the third vein well above the tip of the wing. The diagnostic feature, however, is the loop in the fourth longitudinal vein where it meets the superior cross vein (anterior cross vein) which causes the discal cell to appear 'cleaver-shaped'. The 'angle of rest' is of recognition use, as any brown tropical African fly with a projecting proboscis and wings closely folded over the abdomen like 'closed scissors' is almost certain to be a *Glossina*. The squames as in *Stomoxys* are not particularly large and do not always cover the halteres.

Abdomen—segments of rather leathery texture and seven in number when seen from above (not the usual four of other calypterate flies of medical importance). The first segment is close to the thorax and is almost hidden by the larger second segment. The banding on these segments is of importance in the identification of *Glossina* species.

141. *What are the feeding habits of* Glossina*?*
Glossina of both sexes bite by day. Their compound eyes enable them to see their hosts' movements for a distance of a hundred yards or so. In some instances when it is too hot during the day for the tsetse to move from their resting positions, they have been known to bite at night time, presumably aided by their powers of scent (probably in the antennae). They have also been known to bite by moonlight.

142. *What is the life history of tsetse flies?*
The life history of tsetse is remarkable in that they do not lay eggs or even first stage larvae in the orthodox calypterate fashion, but produce, every ten days or so, a mature third stage larva which has developed within the abdomen of the female. It moults twice within the uterine pouch, and is finally deposited in a selected position by the parent. The larva at this stage is able to wriggle into the loose material on which its is placed, for a period of an hour or so. When buried to its satisfaction it becomes immobile and the last larval skin begins to darken and harden about the pupal cuticle. After some three weeks as a puparium the adult fly emerges with the aid of its ptilinum.

143. *Four species of* Glossina *are used as a basis for explaining the way tsetse species occupy different types of terrain. As these species are* *also associated with certain kinds of hosts and certain kinds of trypanosomes, remark briefly on these associations.*

The riverine areas of West Africa—based mainly on the Niger and its tributaries and the Congo and tributaries, are the habitats of *Glossina palpalis* the main vector of a human type of African trypanosomiasis, designated *Trypanosoma gambiense*, and in the northern hinterland where the rivers tend to dry into pools during the hot season of a smaller species *Glossina tachinoides*. Both these tsetse are closely tied to man as a host and to *Trypanosoma gambiense* as the parasite of sleeping sickness, though *G. palpalis* is not averse to crocodilian or lizard blood in the absence of human beings.

In the game areas of East Africa where there are wide plains and bush country, tsetse species occur which do not require the humid conditions of river-bank vegetation in order to exist. *Glossina morsitans* lives in the bush country, where rocks and fallen trees give it the amount of shade necessary for its survival. *Glossina swynnertoni* is able to live in country without objects of shade other than the bodies of the host animals, which the fly is loath to leave if the sun is hot and the air dry. Both these *Glossina* species are associated with *Trypanosoma rhodesiense* which is the trypanosome strain found in wild game which is infective to man. (*T. brucei* is a game and cattle trypanosome which will cause nagana among domestic animals but never infects man.)

144. *How many species of tsetse are there?*
There are twenty-one species of tsetse, but only a few species are incriminated as vectors of African trypanosomiasis.

145. *What is the supposed route taken by trypanosomes ingested by tsetse and incubated to the infective stage?*
The tsetse fly acquires trypanosomes from the peripheral blood of a host (human or animal) during its blood meal. The alimentary tract of the fly seems to be a suitable environment for the new-comers which enlarge and multiply, producing a large number of slender forms. These multiplied forms pass right down the lumen (endoperitrophic space) of the peritrophic membrane sleeve (which lines the mid-gut and on into the hind-gut for a short distance), and finally by rounding the open ragged end of the

peritrophic sleeve return between the membrane and the gut wall along the ectoperitrophic space. When the slender trypanosomes reach the cul-de-sac of this space at the proventriculus, the parasites pass through the newly formed membrane into the oesophagus. Avoiding incoming meals of blood the trypanosomes pass as soon as possible into the long salivary glands, in which organs they are usually found as shorter crithidial forms (with the flagellum originating *before* the nucleus). However, when the trypanosomes emerge in the saliva which is injected into a new host they have assumed the short stumpy metacyclic forms (with flagellum originating *behind* the nucleus) which may set up a chancre at the site of inoculation. They then proceed to multiply in other parts of the body.

146. *Which are the two cockroach species most likely to be encountered in domestic premises in the United Kingdom?*
Blattella germanica the German cockroach, and *Blatta orientalis* the oriental cockroach.

147. *Compare the appearance of these two species.*
The adult *Blattella germanica* is 12–15 mm. long, yellowish-brown in colour and both sexes have wings.

The adult *Blatta orientalis* is 25–30 mm. long, and dark brown to black in colour. Females have very short vestigial wings covering only the front part of the thorax, and males' wings are slightly larger, extending halfway down the abdomen.

148. *What are the habits of pest cockroaches?*
Pest cockroaches in domestic premises will hide during the day in cracks and crevices, dead spaces, ducts, gulleys, behind fittings etc. and come out at night to feed on any available food, refuse and scraps.

149. *Why are cockroaches a potential health hazard?*
Cockroaches habitually feed by regurgitating digestive fluid from the crop, chewing it into the intended meal and swallowing the mixture. They will readily feed on a variety of human detritus such as faeces, urine, sputum and dustbin refuse from all of which they may take up disease organisms, and will just as readily feed on food intended for human consumption, depositing the organisms

during the feeding process. In addition they will take up organisms on their feet and body surfaces and deposit them on human food and domestic equipment, at the same time defaecating.

150. *What is the sequence of activity in cockroach control?*

1. Maintenance of standards of hygiene and cleanliness in premises.

2. Regular inspection for signs of infestation.

3. Use of residual and knockdown insecticides as both prevention and if necessary control.

4. Monitoring of results of control measures.

5. Regular application as recommended by insecticide manufacturer.

CHAPTER 11

MEMORANDA ON DISEASES TRANSMITTED AND CONDITIONS CAUSED BY ARTHROPODS

The following pages show, in tabular form, the most important anthropod-borne diseases and conditions, their geographical distribution, causative organism, vector and example species, and animal reservoir.

Disease	Cause
DENGUE FEVER Tropical and subtropical regions.	Filter-passing virus in peripheral blood is ingested by vector day before and up to 7th day of fever; 8–11 days incubation in vector. Transmission by bite.
ENCEPHALITIS Mosquito-borne: *St. Louis encephalitis, Eastern, Western* and *Venezuelan equine encephalomyelitis, Japanese B encephalitis.*	Neurotropic filter-passing virus in peripheral blood of animal reservoirs is ingested by vector and transmitted by bite.
Tick-borne: *Russian spring-summer encephalitis* (forest areas), *Colorado tick fever.*	Neurotropic filter-passing virus in peripheral blood of rodent reservoirs, and in body of adult and later larval vectors is transmitted in spring and summer respectively by bite.
EXCREMENTAL DISEASES & OTHER MECHANICALLY TRANSMITTED DISEASES due to pathogenic bacteria, protozoa, helminths, etc.	*Salmonella* and *Shigella spp., Entamoeba histolytica, Vibrio cholerae,* and eggs of certain helminths. Transmission by body hairs, and foot-pads, vomit drop and faeces.

Vectors	Example species	Reservoirs
Four mosquito species (females only).	*Aëdes aegypti* *Aëdes albopictus*	Man, possibly monkeys, wild birds and bats.
Some 5 genera of mosquitoes are implicated (females only).	*Culex pipiens fatigans,* Japanese B encephalitis. *Culex tarsalis,* St. Louis encephalitis.	Small rodents and birds (*Dermanyssus gallinae* mite is transmitter and reservoir of St. Louis encephalitis among chickens).
Possibly 4 Ixodid hard tick species (males and females).	*Ixodes persulcatus* (forest tick), Russian spring-summer encephalitis. *Dermacentor andersoni,* Colorado tick fever.	Ticks (virus over-winters in tick vector). Also small rodents.
Many Calypterate and other flies which visit both infected matter (*e.g.* : human faeces) and human foods. Cockroaches may also be involved in mechanical transmission of these organisms and also bacteria such as *Escherichia coli, Serratia marcescens, Klebsiella spp.,* and *Pseudomonas aeruginosa.*	*Musca domestica* (temperate regions), *Calliphora spp.* *Blattella germanica, Blatta orientalis* and other domestic pest cockroach species in both tropical and temperate regions.	Mainly human excreta, and also surgical swabs, open sores etc.

Disease	Cause
FILARIASIS Tropical and subtropical regions.	*Wuchereria bancrofti* roundworms. Microfilariae (periodic) are in peripheral blood and are infective to vector at night, or both day and night (non-periodic in Fiji and other Pacific Islands). 10–20 days development in vector. Transmission by bite.
Indian, Chinese and Malayan Orient, Polynesian Oceania (S.W.).	*Brugia malayi* roundworms. Microfilariae (periodic) are in peripheral blood, and are infective to vector at night. 6 or more days development in vector. Transmission by bite.
West Africa.	*Loa loa* roundworms. Microfilariae (periodic) in peripheral blood are infective to vector by day; 10–12 days to development in vector. Transmission by bite.
Tropical Africa and Tropical America.	*Onchocerca volvulus* roundworms. Microfilariae (non-periodic) in skin and tissue juices are ingested by vector during the day; 6–7 days development in vector. Transmission by bite.
Tropical Africa and America.	*Dipetalonema perstans* roundworms. Microfilariae (non-periodic) in peripheral blood are ingested by vector during dusk or night; 7–9 days development in vector. Transmission by bite.

Vectors	Example species	Reservoirs
Some 9 *Aëdes* mosquito species, 32 *Anopheles* mosquito species, 14 *Culex* mosquito species, 5 *Mansonia* mosquito species (females only).	*Aëdes scutellaris* (E. Indies and S. Pacific Islands, associated with non-periodic forms) *Anopheles gambiae* (W. Africa). *Culex pipiens fatigans* (tropics and sub-tropics, classically associated with periodic forms). *Mansonia uniformis* (C. Africa).	Man.
Some 7 *Mansonia* mosquito species, some 3 *Anopheles* mosquito species, 1 *Aëdes* mosquito species (females only).	*Mansonia annulata* (Indonesia, Philippines, Malaysia), *Anopheles hyrcanus sinensis* (China), *Aëdes togoi* (China, Japan).	Man.
Some 5 *Chrysops* gadflies (Tabanidae) (females only).	*Chrysops silacea* *Chrysops dimidiata*	Man, monkeys.
Some 5 *Simulium* black-fly species (females only).	*Simulium damnosum* (Africa). *Simulium ochraceum* (America).	Man, possibly other animals.
Some 3 *Culicoides* midge species (females only).	*Culicoides austeni* (Africa).	Man, chimpanzee and gorilla.

209

Disease	Cause
FILARIASIS (continued) Tropical America.	*Mansonella ozzardi* roundworms. Microfilariae (non-periodic) in peripheral blood are ingested by vector during dusk or night; 6–8 days development in vector. Transmission by bite.
LEISHMANIASIS Tropical America and Africa; Mediterranean; India and Chinese Orient.	*Leishmania donovani,* cause of Kala-azar or visceral leishmaniasis. Leishmanial forms (Leishman Donovan bodies) in peripheral blood and skin are ingested by vector during dusk or night; 7 days development in vector. Transmission by bite.
Mediterranean, N. Africa, Near and Middle East, Tropical America.	*Leishmania tropica,* cause of Oriental sore or cutaneous leishmaniasis. *Leishmania* from area around sores are ingested by vector during dusk or night; believed 7 days development in vector. Transmission by bite.
Tropical America	*Leishmania brasiliensis,* cause of Espundia or muco-cutaneous leishmaniasis. *Leishmania* from area around sores are ingested by vector during dusk or night; believed 7 days development in vector. Transmission by bite.
MALARIA Tropical and temperate regions.	*Plasmodium vivax, Plasmodium falciparum, Plasmodium malariae, Plasmodium ovale.* Sexual forms (gametes) in peripheral blood are ingested by vector at night, at dusk or during the day; 12 days incubation in vector. Transmission by bite.

Vectors	Example species	Reservoirs
Possibly 2 *Culicoides* midge species (females only).	*Culicoides furens* (C. and S. America).	Man.
Some 7 *Phlebotomus* sandfly species (females only).	*Phlebotomus argentipes* (India).	Man, dogs.
Some 3 *Phlebotomus* sandfly species (females only).	*Phlebotomus sergenti* (Middle East) *Phlebotomus papatasii* (N. Africa, E. Mediterranean).	Man, dogs, cats and some other animals.
Possibly 5 *Phlebotomus* sandfly species (females only).	*Phlebotomus intermedius* (C. and S. America) (chiefly forest areas).	Man, some rodents.
Possibly over 100 *Anopheles* mosquito species, some 20 of which are notorious (females only).	*Anopheles gambiae* (Africa) *Anopheles stephensi* (India). *Anopheles maculipennis* (most warm and temperate regions). *Anopheles darlingi* (South America).	Man.

Disease	Cause
MITE DERMATITIS Most areas of the World.	Usually allergic irritation due to bites or burrowing in skin by various mite species.
MYIASIS *Specific myiasis*, mainly in tropical regions.	Larvae (maggots) of Calypterate flies which normally develop only in living tissue.
Semi-specific (wound) myiasis. Tropical and temperate regions.	Larvae (maggots) of Calypterate flies which normally develop in decaying animal or vegetable matter, but which may invade sores or wounds
Accidental myiasis Tropical and temperate regions.	Any stage (usually larval) of some Calypterate and Acalypterate flies, which invade alimentary tract (usually in food) or urogenital passage.
PEDICULOSIS World-wide.	Irritation of skin by multiple bites of human lice. Aided by secondary infection.
PLAGUE *Urban epidemic plague*. Mainly tropical and sub-tropical regions.	*Yersinia pestis* in peripheral blood of rats (which it kills) is ingested by flea vector and transmitted in regurgitated colonies of bacilli from blocked proventriculus by bite of flea.

Vectors	Example species	Reservoirs
Some Sarcoptid, Parasitid, Trombiculid, Tarsonemid, Tyroglyphid and other mites.	*Sarcoptes scabiei* (World-wide). *Pediculodes* (= *Pyemotes*) *ventricosus* grain itch mite. *Cheyletiella spp.* from domestic animals.	Man, other animals; stored food products, etc.
Various Calliphorid and Oestrid larvae.	*Dermatobia hominis* bot fly (Central and South America). *Cordylobia anthropophaga* (tropical Africa).	Living tissue of animals, including Man.
Various Calliphorid larvae.	*Lucilia sericata, Sarcophaga barbata* (widespread).	Carcasses, excreta, etc., and living tissue.
Various Calliphorid, Muscid, Syrphid and other flies.	*Megaselia scalaris* (all stages have occurred in the human gut).	Decaying animal and vegetable matter.
Pediculus humanus capitis and *P.h. corporis* (human head and body lice) and *Phthirus pubis* (pubic or crab louse) (World-wide).		Man only.
All fleas are potential vectors. Some four *Xenopsylla* flea species are most important (males and females).	*Xenopsylla cheopis.*	Brown and black rats.

Disease	Cause
PLAGUE (continued) *Sylvatic plague.* Rural N. America, Central Asia and S. Africa.	*Yersinia pestis* in peripheral blood of wild rodents (less affected than rats), sometimes rabbits, is ingested by vector and occasionally transmitted to man by bite. Blood of sick rodents is infective through abrasions.
RELAPSING FEVER *Louse-borne, epidemic.* World-wide except Australia.	*Borrelia recurrentis* in peripheral blood during febrile periods is ingested by vector and retained in haemocoel from 8 hours after infection. Infection is from crushed body of vector.
Tick-borne, endemic. Warmer areas of Europe, Africa, Asia and America.	*Borrelia duttoni* (African) or other tick-specific strain in peripheral blood of man during febrile periods, or in reservoir animal, is ingested by vector mainly at night. Infection by bite or coxal fluid of vector.
RICKETTSIAL FEVERS *Classical exanthematic epidemic,* or *louse-borne typhus.* World-wide.	*Rickettsia prowazeki* in peripheral blood during febrile period, 3rd–10th day; 5–9 days incubation in vector. Louse faeces main source of infection.
Murine, endemic or *flea-borne typhus.* N. & S. America, Mediterranean region, Africa, Middle and Far East	*Rickettsia mooseri* in peripheral blood of rats and mice (little effect on rodents) is ingested by vectors and transmitted to man sporadically by bite.
Scrub or *mite-borne typhus; Japanese river fever.* Indian, Malayan and Chinese Orient; some Polynesian islands and Australia.	*Rickettsia orientalis* (= R. *tsutsugamushi*) in peripheral blood of rodent reservoir is ingested by larval vector. Later generations of larvae transmit infection by bite.

Vectors	Example species	Reservoirs
Many wild rodent flea species (males and females).	*Oropsylla silantiewi* tarabagan flea (Mongolia).	Wild rodents, *e.g.*: tarabagan (Mongolia), gerbil (S. Africa), ground squirrel (N. America), etc.
Both sexes of the human louse, particularly the body louse *Pediculus humanus corporis*; *P.h. capitis* may also be involved, but not *Phthirus pubis*.		Man.
Some 10 *Ornithodorus* soft tick species.	*Ornithodorus moubata* (Africa).	Man, rodents, other animals and *Ornithodorus* ticks themselves by hereditary transmission.
Both sexes of the human louse, particularly the body louse *Pediculus humanus corporis*. *P.h. capitis* may also be involved but not *Phthirus pubis*.		Man.
Some 5 genera of fleas; also rat louse *Polyplax spinulosa*, and tropical rat-mite *Ornithonyssus bacoti* (hereditary infection).	*Xenopsylla cheopis* (USA and elsewhere)	Brown and black rats, mice, tropical rat-mite *Liponyssus bacoti*.
Some four *Trombicula* mite species (only larval forms are parasitic and transmit disease to man).	*Trombicula deliensis* (Far East), *Trombicula akamushi* (Japan).	Rodents including rats, and nymphal and adult *Trombicula* mites (hereditary transmission).

Disease	Cause
RICKETTSIAL FEVERS (continued) *Tick-borne typhus* (including American spotted fevers).	*Rickettsia rickettsi* in infected vector at any stage or in peripheral blood of animal reservoir, is ingested by vector and sporadically transmitted to man by bite.
Boutoneuse fever. Mediterranean (probably identical with Kenya tick typhus spotted fevers).	*Rickettsia rickettsi* var. *conori* in infected vector tick at any stage, or in peripheral blood of animal reservoir, is ingested by vector and sporadically transmitted to man by bite.
South African tick-bite fever. S. Africa, S.E. Asia.	*Rickettsia rickettsi* var. *pijperi* in infected vector tick is sporadically transmitted to man by bite.
'Q' Fever Australia, North America, Europe.	*Coxiella burneti* (= *Rickettsia diaporica*) in peripheral blood of reservoir animals, is ingested by tick vectors. Transmission by faeces of tick vectors, either in bites or wounds, but most commonly by respiratory invasion.
SANDFLY FEVER Mediterranean, Indian and Chinese Orient, E. Africa and S. America.	Filter-passing virus in peripheral blood, infective to vector the day before and up to 4th day of fever. 6–9 days incubation in vector. Transmission by bite.
TUNGIASIS Tropical America, Africa, including Madagascar (dry, sandy areas). Also parts of India and Far East.	Female jigger flea burrows in living tissue (usually feet), leading to secondary infection.

Vectors	Example species	Reservoirs
Possibly 7 Ixodid hard tick species.	*Dermacentor andersoni* (Western USA), *Dermacentor variabilis* (Eastern USA), *Amblyomma cajennense* (Brazil and Colombia), *Amblyomma americanum* (Texas and other areas).	Vector ticks, also strictly zoophilic ticks, small rodents (not large mammals) and occasionally man.
One Ixodid hard-tick species.	*Rhipicephalus sanguineus* dog tick (S. Europe and Africa).	Dogs (most important), some rodents, and vector ticks (hereditary).
Possibly 3 Ixodid hard tick species.	*Haemaphysalis leachi* dog-tick, *Ambylomma hebraeum*.	Dog tick *Haemaphysalis leachi*.
Possibly 8 Ixodid hard tick species and one Argasid soft tick species.	*Ixodes holocyclus* (bites bandicoots, man and other animals) (Australia). *Dermacentor andersoni* (N. America).	Bandicoot and other bush animals (Australia). *Haemaphysalis humerosa* (usually bandicoot specific). In other areas reservoirs unknown.
One or more *Phlebotomus* sandfly species (females only).	*Phlebotomus papatasii*.	Man.
Tunga penetrans jigger flea.	*Tunga penetrans*.	Man, domestic animals including fowl

Disease	Cause
TRYPANOSOMIASIS Western and Central Tropical Africa. (epidemic forms.)	*Trypanosoma gambiense* in peripheral blood of man (or reservoir animal) is ingested by vector; 18–20 or more days incubation. Transmission by bite.
East and Central Tropical Africa (limited and sporadic form).	*Trypanosoma rhodesiense* in peripheral blood of reservoir animal (or man), is ingested by vector; 18–20 or more days incubation in vector. Transmission by bite.
Tropical America.	*Trypanosoma cruzi* in peripheral blood of man and other animals is ingested by vector and transmitted to man in bug faeces while vector is feeding; 6–15 days incubation in vector.
TULARAEMIA USA, Canada, Japan, Russia, Central Europe, Tunisia.	*Francisella tularensis* in peripheral blood of rodent reservoirs is ingested by vector. Bites, crushed bodies and faeces of vectors, and tissues and body fluids of rodent reservoirs, are highly infective (also water of some natural streams and lakes).
YELLOW FEVER Widely distributed in America and Africa. Endemic centres in virgin forest.	Filter-passing virus in peripheral blood during first 3–5 days after onset; 9–14 days incubation in vector. Transmission by bite.

Vectors	Example species	Reservoirs
Some 5 *Glossina* tsetse fly species (males and females).	*Glossina palpalis* (wet riverine species), *Glossina tachinoides* (dry riverine species).	Man (most important); bush and marsh game and domestic animals are potential reservoirs.
Some 3 *Glossina* tsetse fly species (males and females).	*Glossina morsitans* (open country species). *Glossina swynnertoni* (dry open country species).	Wild game and domestic farm animals (most important), also man.
Many Reduviid (Triatomine) cone-nose bug species in some 8 genera.	*Panstrongylus megistus, Rhodnius prolixus, Triatoma infestans.*	Man, dogs, cats, bats, rodents, armadilloes, opossums etc.
Many arthropod species, including ticks, mosquitoes, Tabanid flies, and fleas.	*Chrysops discalis* gadfly (USA).	Various wild rodents, *e.g.* cotton-tail rabbits, hares, squirrels, etc. Also some birds, carnivores and ungulates.
One *Aëdes* mosquito species (urban epidemics). Possibly 6 or more genera of mosquitoes are implicated in jungle yellow fever (females only).	*Aëdes aegypti* (urban yellow fever). *Haemogogus capricorni, Aëdes simpsoni* (jungle yellow fever).	Man in urban areas, jungle animals, mainly monkeys, in endemic forest centres.

219

INDEX